Wireless Networks

Series Editor

Xuemin Sherman Shen, University of Waterloo, Waterloo, Canada

The purpose of Springer's Wireless Networks book series is to establish the state of the art and set the course for future research and development in wireless communications networks. The scope of this series includes not only all aspects of wireless networks (including cellular networks, WiFi, sensor networks, and vehicular networks), but related areas such as cloud computing and big data. The series serves as a central source of references for wireless networks research and development. It aims to publish thorough and cohesive overviews on specific topics in wireless networks, as well as works that are larger in scope than survey articles and that contain more detailed background information. The series also provides coverage of advanced and timely topics worthy of monographs, contributed volumes, textbooks and handbooks.

List of Figures

Tho Le-Ngoc • MohammadMahdi Ghadaksaz •
Mobeen Mahmood

AI-Enabled UAV-Assisted Massive MIMO

Relaying for 5G-and-Beyond Wireless Networks

Springer

Tho Le-Ngoc
Department of Electrical and Computer Engineering
McGill University
Montreal, QC, Canada

MohammadMahdi Ghadaksaz
Department of Electrical and Computer Engineering
McGill University
Montreal, QC, Canada

Mobeen Mahmood
Department of Electrical and Computer Engineering
McGill University
Montreal, QC, Canada

ISSN 2366-1186 ISSN 2366-1445 (electronic)
Wireless Networks
ISBN 978-3-032-07871-1 ISBN 978-3-032-07872-8 (eBook)
https://doi.org/10.1007/978-3-032-07872-8

This Springer imprint is published by the registered company Springer Nature Switzerland AG
The registered company address is: Gewerbestrasse 11, 6330 Cham, Switzerland

Preface

The International Telecommunication Union (ITU)'s 2030 sixth generation (6G) framework focuses on enhancing terrestrial networks by integrating aerial communications for peak data rate, ubiquitous coverage, and sensing. Direct long-distance links, however, often fall short of these targets or lead to excessive power consumption, while static base stations (BSs) struggle to adapt to dynamic environments and are costly to deploy in urban areas. As a solution, relay-based communications offers a more efficient alternative. In this context, unmanned aerial vehicles (UAVs) have emerged as a promising mobile, aerial relay solution in beyond fifth generation (5G) mobile networks. Their ability to reposition dynamically allows for optimized network coverage, making them ideal for scenarios like emergency response, network offloading, periodic sensing, and large data transfers in 5G and beyond (5G&B) networks. Equipping such UAV relays with large-scale massive MIMO (mMIMO) arrays and hybrid beamforming (HBF) allows them to harness the millimeter-wave (mmWave) band, meeting the stringent throughput and latency requirements expected of future networks.

In this book, we investigate various UAV-assisted mMIMO relaying scenarios, where the direct transmission from a BS to a gateway/several ground users is non-feasible due to potential blockage. We aim to maximize the total achievable rate (AR) while considering energy-efficient configurations. Specifically, we propose different Artificial Intelligence (AI)-based solutions as robust and low-complexity approaches for the complex problems explored in this book.

First, we investigate particle swarm optimization (PSO)-based algorithmic solutions for UAV-assisted mMIMO AR maximization problems. We begin with a single-user mMIMO (SU-mMIMO) scenario in which the UAV acts as a relay between a BS and gateway. Considering various relaying schemes and beamforming setups for the UAV, we propose a PSO-based UAV location optimization approach to maximize the total AR between the BS and the gateway. Subsequently, we extend the problem to a multi-user mMIMO (MU-mMIMO) setting that replaces the gateway with multiple ground users. Here, we propose a joint HBF, UAV location and power allocation (PA) optimization to achieve a near optimal sum-rate. Consequently, we further expand the approach to multiple-UAV MU-mMIMO cooperative relaying scenario to further improve the capacity and coverage. The illustrative results demonstrate the effectiveness of the proposed approaches to accomplish near-optimal sum-rate and energy-efficient performance while significantly extending the overall service area.

Second, we explore several deep learning (DL)-based solutions to reduce the computational complexity of the PSO-based solutions without compromising performance. Here, we examine the UAV-assisted MU-mMIMO scenario with the BS, UAV, and several ground users where the only means of communications between the users and the BS is the UAV. Considering HBF and the UAV acting as a decode-and-forward (DF) relay, we aim to maximize the total AR. Initially, we propose a deep supervised learning (DSL)-based UAV location and PA optimization to reduce the runtime of the previously used PSO-based solution. Next, we extend the problem to scenarios in which users can change their location and then propose a deep reinforcement learning (DRL)-based UAV deployment in dynamic environments. Finally, we develop a DRL codebook-based UAV deployment and trajectory design for multiple-UAV scenarios. The numerical results reveal that the proposed algorithms can significantly reduce the computational complexity of the PSO-based solutions while maintaining a satisfactory AR.

Montréal, QC, Canada — Tho Le-Ngoc
Montréal, QC, Canada — MohammadMahdi Ghadaksaz
Ottawa, ON, Canada — Mobeen Mahmood
August 2025

Declarations

Competing Interests The authors have no competing interests to declare that are relevant to the content of this manuscript.

Contents

List of Tables

Acronyms

1-D	One-dimensional
2-D	Two-dimensional
3-D	Three-dimensional
3GPP	Third-generation partnership project
5G	Fifth generation
5G&B	5G and beyond
6G	Sixth generation
AAoA	Azimuth angle of arrival
AAoD	Azimuth angle of departure
ABF	Analog beamforming
ACO	Ant colony optimization
AF	Amplify-and-forward
AI	Artificial intelligence
AoA	Angle of arrival
AoD	Angle of departure
AR	Achievable rate
BB	Baseband
bps	Bits per second
BS	Base station
CM	Constant modulus
CDRL	Constrained DRL
CNN	Convolutional neural network
CSI	Channel state information
DDPG	Deep deterministic policy gradient
DDPG-C-UD	DDPG codebook-based UAV deployment
DDPG-UD	DDPG-based UAV deployment
DDQN	Double DQN
DF	Decode-and-forward
DL	Deep learning
DL-UD	DL-based UAV deployment
DNN	Deep neural network

DoF	Degrees of freedom
DQN	Deep Q-network
DRL	Deep RL
DSL	Deep SL
dB	Decibel
dBm	Decibel milliwatts
EAoA	Elevation angle of arrival
EAoD	Elevation angle of departure
EB	Exabytes
EH	Energy harvesting
eMBB	Enhanced mobile broadband
FDBF	Fully digital beamforming
FD	Fixed UAV deployment
FIFO	First-in-first-out
GA	Genetic algorithm
GHz	Gigahertz
HBF	Hybrid beamforming
HD	Half-duplex
HL	Hidden layer
Hz	Hertz
IWIPT	Integrated wireless information and power transfer
ITU	International Telecommunication Union
IoT	Internet of things
IUT	Inter-user interference
JGP	Joint-group-processing
J-HBF-DSLLPA	Joint HBF, DSL-based UAV location and PA
J-HBF-FL-EQPA	Joint HBF, fixed UAV location and equal PA
J-HBF-PSOPA-FL	Joint HBF, PSO-based PA and fixed UAV location
J-HBF-PSOLPA	Joint HBF, PSO-based UAV location and PA
J-HBF-PSOL-EQPA	Joint HBF, PSO-based UAV location and equal PA
JSAC	Joint sensing and communications
LTE	Long-term evolution
M2M	Machine-to-machine
MAE	Mean absolute error
MARL	Multi-agent reinforcement learning (RL)
MF	Matched filter
MIMO	Multiple-input multiple-output
ML	Machine learning
mMIMO	Massive MIMO
mMTC	Massive machine type communications
mmWave	Millimeter-wave
MSE	Mean square error
MU	Multi-user
MU-MIMO	Multi-user MIMO
MU-mMIMO	Multi-user mMIMO

NI	Nature-inspired
NLP	Natural language processing
NOMA	Non-orthogonal multiple access
NP-hard	Non-deterministic polynomial-time hardness
NLoS	Non-LoS
LoS	Line-of-sight
PA	Power allocation
PPO2	Proximal policy optimization 2
PSO	Particle swarm optimization
PSO-UD	PSO-based UAV deployment
PS	Phase shifter
QoS	Quality of service
RF	Radio frequency
ReLU	Rectified linear unit
RL	Reinforcement learning
RNN	Recurrent neural network
RZF	Regularized zero-forcing
SINR	Signal-to-interference-plus-noise ratio
SL	Supervised learning
SU-MIMO	Single-user MIMO
SVD	Singular value decomposition
THz	Terahertz
UAV	Unmanned aerial vehicle
UE	User equipment
UL	Unsupervised learning
ULA	Uniform linear array
URA	Uniform rectangular array
URLLC	Ultra reliable low latency communications
V2N	Vehicle-to-network
V2P	Vehicle-to-pedestrian
V2V	Vehicle-to-vehicle
V2X	Vehicle-to-everything
WIT	Wireless information transfer
WPT	Wireless power transfer
WSN	Wireless sensor network
ZB	Zettabytes
ZF	Zero-forcing

Symbols

α	Actor network learning rate
$\mathbf{a}$	Action vector
$\mathbf{B}, \mathbf{b}$	Digital BB precoder matrix and vector
β	Critic network learning rate
$\mathbf{d}$	Data signal vector
$\delta^{\theta}, \delta^{\psi}$	Elevation angular, azimuth angular spread
D	Replay buffer size
d	Antenna spacing normalized by the wavelength
$\mathbf{e}$	Steering vector
$\mathbf{F}, \mathbf{f}$	RF beamformer matrix and vector
G	Number of user groups
g	User group index
γ	Discount factor
$\mathcal{H}$	Effective channel matrix
$\mathbf{H}, \mathbf{h}$	Channel matrix and vector
η	Path loss exponent
K	Total number of users
K_g	Number of users in the gth user group
λ^x, λ^y	Quantized angle-pairs
$\mathcal{L}$	Loss function
L	Total number of paths
μ, μ'	Actor and target actor networks
N_{RF}	Number of RF chains
N_r	Number of antennas at UAV receive
N_t	Number of antennas at UAV transmit
N_T	Number of antennas at the BS
ν	Target networks learning rate
$\mathbf{n}$	Noise vector
Ω_0	Reward threshold
P_T	Transmit power
$\mathbf{P}$	Power allocation matrix

p_k	Allocated power to the kth user
Q, Q'	Critic and target critic networks
R_{sum}	Achieved sum-rate
r	Scalar reward
$\mathbf{s}$	State vector
S_r	Dataset size
T	Number of iterations
θ, ψ	Mean azimuth AoD, mean elevation AoD
$\mathbf{W}$	Neural network's weight matrix
x, y, z	3-D location coordinates
ξ	RZF regularization parameter
$\mathbf{Z}$, $\mathbf{z}$	Path gain matrix and vector
ζ_i	Number of neurons at ith hidden layer

Notations

- Bold upper/lower case letters denote matrices/vectors.
- $\mathbb{C}$ and $\mathbb{R}$ represent the set of complex and real numbers, respectively.
- $(\cdot)^*$, $(\cdot)^T$, $(\cdot)^H$, $\|\cdot\|_1$, $\|\cdot\|_2$, and $\|\cdot\|_F$ are the complex conjugate, the transpose, the conjugate transpose, the 1-norm, the 2-norm, and the Frobenius norm, respectively.
- $\mathbf{I}_K$, $\mathbb{E}\{\cdot\}$, $\mathrm{tr}(\cdot)$, and $\angle(\cdot)$ stand for $K \times K$ identity matrix, the expectation operator, the trace operator, and the argument of a complex number, respectively.
- $\mathbf{X}[m, n]$ denotes the element at the intersection of mth row and nth column.
- $\mathbf{X} \otimes \mathbf{Y}$ is the Kronecker product of two matrices $\mathbf{X}$ and $\mathbf{Y}$.
- $\mathrm{sgn}(x) = \begin{cases} +1, & x > 0, \\ 0, & x = 0, \\ -1, & x < 0 \end{cases}$ denotes the sign function.
- $\mathbf{X} \odot \mathbf{Y}$ is the Hadamard product of two matrices $\mathbf{X}$ and $\mathbf{Y}$.
- $\mathrm{Span}(\mathbf{X})$ denotes the space spanned by the columns of $\mathbf{X}$.
- $\mathrm{Null}(\mathbf{X})$ denotes the null space of $\mathbf{X}$ with the set of all vectors $\mathbf{y}$ such that $\mathbf{Xy} = \mathbf{0}$.
- $\mathrm{diag}(\mathbf{x})$ creates a diagonal matrix with the elements of a given vector $\mathbf{x}$.
- $\mathrm{clip}\,(x, [a, b]) = \max\,(a, \min\,(x, b))$ denotes a clipping function with $a \leq b$.
- $(x)^+ = \max(0, x)$ refers the ramp function.
- $x \sim \mathcal{CN}(0, \sigma^2)$ denotes a complex Gaussian random variable with zero-mean and variance σ^2.

Chapter 1
Introduction

1.1 Overview of Fifth Generation and Beyond

Over the years, the evolution of cellular networks from the first generation (1G) to the fifth generation (5G) has had a profound impact on different aspects of our lives. Particularly, the world has seen a rapid digital transformation in the last few years that has changed the way people communicate, conduct business, and search for information. A critical element of this digital transition is wireless connectivity [1, 2]. It is expected that 5G and beyond (5G&B) networks will pave the way toward realizing the individuals technological aspirations including holographic telepresence, e-health, pervasive connectivity in smart environments, massive robotics, three-dimensional (3-D) massive unmanned mobility, augmented reality, virtual reality, and Internet of Everything [3]. In this regard, the following statistics highlight the significant increase in data traffic in the next few years:

- According to the International Telecommunication Union (ITU), the annual data traffic increase between 2020 and 2030 is estimated to be around 55%, which can reach 606 exabytes (EB) per month in 2025 and 5016 EB per month in 2030 as shown in Fig. 1.1 [4].
- It is expected that the number of smart devices will surpass 38.9 billion by 2029 (an increase of 148 % when compared to 15.7 billion devices in 2023) [5, 6].
- As per ITU, the overall mobile data traffic will reach 5 zettabytes (ZB) per month [7].

The explosive increase of mobile data traffic is real, driven primarily due to the increased use of smart phones, tablets, video streaming services as well as machine-to-machine (M2M) connections. Hence, with billions of low-power consumption devices, such as wireless sensors, connecting to the Internet through the Internet of Things (IoT) framework, there is a significant increase in both energy demand and data traffic. Hence, the need for a new generation of highly scalable cellular networks is inescapable. In particular, this new generation should:

T. Le-Ngoc et al., *AI-Enabled UAV-Assisted Massive MIMO*, Wireless Networks,
https://doi.org/10.1007/978-3-032-07872-8_1

ence. Additionally, exploring new frequency spectrums, such as millimeter-wave (mmWave) communications and sub-terahertz (THz) (sub-THz) communications, can provide higher bandwidth.

1.2 UAV-Assisted Communications

With the advent of advanced wireless communications and networking technologies, a new era of innovation has emerged across a wide range of applications. These technologies hold transformative potential in sectors such as healthcare, urban development, and everyday environments [15]. However, realizing their full benefits requires overcoming significant challenges, particularly in achieving efficient information transfer between wireless nodes and gateways. To address this issue, various routing schemes have been proposed, including direct transmission or relay structures. Nonetheless, when the distance between the end node and the gateway is substantial, direct transmission may not be feasible or may consume excessive power. In such cases, communications through relay can be a more power-efficient alternative. Moreover, deploying cellular stations in urban areas can be a costly and challenging task, which can further complicate the communications coverage issues in these frameworks [16]. UAVs, commonly referred to as drones, are viewed as a key component of the next generation of wireless communications networks. UAV as a relay offers several advantages over traditional static relays. Specifically, the ability to deploy on-demand, mobile relaying systems at a relatively low cost and in a timely manner makes them particularly well-suited for unforeseen or short-term events, such as emergency situations or network offloading [17]. Furthermore, the high mobility of UAVs allows for the dynamic adjustment of their locations to optimize communications conditions, a technique particularly promising for delay-tolerant applications, such as periodic sensing and the transfer of large data [18–20]. UAVs' capability to reach inaccessible locations makes them a viable option for future IoT applications, as they can fly close to IoT devices, sequentially collect sensing data, address coverage issues, and reduce IoT communications networks' overhead [21].

The incorporation of UAVs as relay nodes in wireless sensor networks (WSNs) has the potential to augment communications capacity by connecting remote sensor gateways and addressing the escalating data rate demands in applications such as virtual reality, device-to-device communications, and smart cities. UAVs can be deployed at high altitudes to increase the likelihood of LoS dominated air-to-ground communications channels, thereby supporting high-rate communications. However, the severely congested sub-6 gigahertz (GHz) bands can be inadequate to meet the rising data rate requirements. In contrast, mmWave communications, with their abundant spectrum resources, can potentially support the high-throughput and low-latency requirements of various UAV application scenarios [22]. Nonetheless, mmWave signals suffer from high propagation loss, including free-space path loss, atmospheric and molecular absorption, and rain attenuation. This challenge can be surmounted by leveraging mMIMO technology with large array structures to generate high beam gains in UAV-assisted terrestrial systems, which can improve the transmission range and simultaneously suppress interference among users/devices by utilizing the advanced capabilities of 3-D beamforming.

1.3 MIMO to Massive MIMO (mMIMO): A Paradigm Shift in Wireless Technology

multiple-input multiple-output (MIMO) technology revolutionizes wireless communication by deploying multiple antennas at both the transmitter and receiver ends. This approach significantly enhances capacity and performance by mitigating challenges like path loss and multipath fading that typically degrade wireless signals. Unlike traditional single-antenna systems, MIMO capitalizes on multipath propagation, exploiting the various channel conditions between each antenna pair to provide additional degrees of freedom (DoFs). MIMO technology is designed to deliver two key benefits: (i) spatial multiplexing gain, which allows the transmission of parallel data streams without the need for extra bandwidth, thereby boosting spectral efficiency and data rates, and (ii) diversity gain, which leverages distinct channel paths to improve signal quality, reducing errors and enhancing reliability. However, a balance must be struck between these benefits, known as the diversity-multiplexing trade-off. Over the past 15 years, MIMO has become a cornerstone of modern wireless standards, including long-term evolution (LTE) and WiFi [23–25]. For example, LTE-Advanced supports configurations up to 8×8 MIMO, showcasing its integral role in advancing wireless communications technology [26].

Massive MIMO represents a significant evolution in MIMO technology by addressing the need to serve multiple users simultaneously with enhanced efficiency. Unlike traditional MIMO systems, which primarily focused on point-to-point communication between two devices with multiple antennas, mMIMO is designed for multi-user mMIMO (MU-mMIMO) scenarios [27]. Here, a transmitter (e.g., a base station (BS) or UAV) equipped with a large array of antennas can simultaneously serve many users, each with a single antenna. This allows for the multiplexing gain to be shared among all users, significantly boosting spectral efficiency and network capacity. The large number of antennas in mMIMO systems also enables the use of simple linear signal processing techniques, to efficiently manage interference and optimize performance across all users, making mMIMO a cornerstone technology for 5G&B. The 3GPP has recognized the potential of mMIMO, standardizing deployments of up to 256 antennas at BS in its Release 17 [9].

1.3.1 Benefits

The deployment of large antenna arrays and the capability for 3-D beamforming in mMIMO systems offer several key benefits:

- **High Spectral Efficiency:** In MU-mMIMO systems, large antenna arrays allow a BS/UAV to serve multiple users simultaneously on the same frequency band. Consequently, mMIMO technology enhances both spectral efficiency and overall network capacity.
- **Improved Interference Management:** The narrow beams generated by large antenna arrays in mMIMO systems significantly reduce inter-user interference (IUI). Advanced beamforming techniques not only enhance the desired signal power but also suppress interference power.
- **High Energy Efficiency:** Large antenna arrays in mMIMO systems reduce downlink and uplink transmit powers through advanced beamforming techniques. This enhancement in array gains leads to higher energy efficiency, which is crucial for certain mMTC applications.
- **Scalability:** mMIMO systems are highly scalable, making them well-suited for future expansions in wireless networks. This scalability ensures that as the demand for data and connectivity grows, mMIMO systems can be adjusted to meet these increasing needs without substantial changes to the existing infrastructure.
- **Flexibility in Deployment:** The advanced beamforming capabilities of mMIMO systems provide greater flexibility in deployment scenarios. This indicates that the mMIMO can also be applicable to UAV-assisted communications or other emerging technologies (e.g., satellite communications).

1.3.2 Challenges

Although mMIMO technology offers substantial benefits, its practical implementation in 5G&B wireless networks brings several technical challenges. Key challenges include:

- **High Hardware Costs and Complexity:** One of the most significant obstacles is the high cost and complexity of hardware. Traditional MIMO systems typically rely on fully digital beamforming (FDBF), where each antenna element is associated with its own radio frequency (RF) chain, digital-to-analog converters (DACs) for transmission, analog-to-digital converters (ADCs) for reception, mixers, and other components. In mMIMO systems, the large number of antenna elements dramatically increases hardware costs and power consumption. While substantial beamforming gains can reduce transmit power without compromising spectral efficiency, the high power consumption of these hardware components can significantly impact energy efficiency. This can be even more challenging for UAV-related applications with limited power resources.

- **Substantial Channel Estimation Overhead:** In mMIMO systems using traditional FDBF techniques, estimating the high-dimensional channel state information (CSI) is crucial for effective beamforming. The increased number of antennas leads to higher channel estimation overhead. Given a fixed coherence time (the period during which the channel remains stable), acquiring extensive CSI requires longer pilot transmission times, which in turn reduces the time available for data transmission. This increased overhead can negatively affect effective spectral efficiency. For example, as the number of antennas approaches infinity, the required pilot transmission length could surpass any finite coherence time, resulting in zero effective spectral efficiency. Therefore, it is essential to develop new beamforming methods that require less CSI overhead for mMIMO systems.
- **High Computational Complexity:** As the size of the antenna array increases, so does the size of the channel matrix. Applying conventional MIMO signal processing algorithms to tasks such as beamforming, channel estimation, and signal detection involves large-scale matrix computations (e.g., matrix inversions, multiplications, singular value decomposition (SVD), and determinants), leading to higher computational complexity.
- **Dynamic and Unpredictable Channel Conditions:** The integration of UAVs introduces highly dynamic and rapidly changing channel conditions due to their mobility and varying altitudes. This makes channel estimation and beamforming in mMIMO systems more challenging, as traditional techniques may not adapt quickly enough to these variations [21].

The advantages of mMIMO technology significantly enhance 5G and future networks, enabling support for emerging services, particularly eMBB and mMTC applications. However, to fully exploit these benefits in UAV-assisted terrestrial networks, it is vital to address the technical challenges. This necessitates the development of advanced, low-complexity signal processing techniques to reduce hardware costs, complexity, and channel estimation overhead.

Hybrid beamforming (HBF) has been proposed as a promising solution for mMIMO systems [28]. Unlike the conventional single-stage FDBF, HBF divides the beamforming architecture into two stages: an analog RF stage and a digital baseband (BB) stage, interconnected by a reduced number of RF chains. This HBF architecture helps mitigate the technical challenges of high hardware cost and complexity, large CSI overhead, and increased computational demands. A detailed comparison of beamforming architectures will be presented in Sect. 2.2.

1.4 The Rise of Artificial Intelligence (AI)

Earlier in this chapter, we discussed that in 5G&B, we will witness a vast variety of emerging applications such as IoT, V2X, and M2M that can directly affect our everyday life where:

- **IoT:** IoT refers to a network of physical devices, vehicles, appliances, and other physical objects that are embedded with sensors, software, and network connectivity, allowing them to collect and share data. The applications of IoT are vast, ranging from healthcare and urban environments (e.g., smart transportation) to households (e.g., smart homes) [15].
- **V2X:** V2X refers to a unified communications framework enabling vehicles to exchange information with their surrounding environment, including vehicle-to-vehicle (V2V), vehicle-to-pedestrian (V2P), and vehicle-to-network (V2N) [29].
- **M2M:** Communication focuses on direct interactions between machines without human intervention, which is a key enabler for industrial automation, intelligent manufacturing, and remote monitoring systems [30].

The aforementioned applications will further force the wireless systems to meet the eMBB, mMTC, and URLLC requirements. Furthermore, as mentioned earlier, the wireless systems for 5G&B must be able to support a tremendous number of users and higher data rates, even under varying channel and user conditions [31]. In this case, conventional algorithms and solutions are expected to be no longer sufficient to meet these requirements.

With the success of AI algorithms in other fields such as natural language processing (NLP), computer vision, and speech recognition, researchers have been motivated to consider using AI techniques in wireless communications. AI is a multidisciplinary technology that consists of multiple research areas such as machine learning (ML), nature-inspired (NI) algorithms, expert systems, etc. [32]. AI technologies will not only reduce or eliminate manual efforts involved in network design, configuration, and management, but also enhance the performance, reliability, and adaptability of communications networks by enabling robust real-time decisions based on predictions of network and user behavior [33].

The applications of AI in wireless communications are extensive, although the field is still in its early stages. From channel estimation to resource allocation, AI has shown satisfactory performance in solving the problems. Motivated by this, we are considering AI-enabled solutions for the UAV-assisted mMIMO problems in this book. In particular, it is important to use algorithmic solutions that can handle the large number of users enabled by using mMIMO configuration and adapt the solution to dynamic scenarios in UAV-assisted communications. Furthermore, it is important to use solutions that have low computational complexity, while achieving near-optimal performance. Due to the aforementioned benefits, AI will be suitable to not only solve the non-convex, multidimensional UAV-assisted mMIMO problems here, but also keep the computational complexity low. We will discuss our AI-based solutions in detail in the following chapters.

1.5 Book Organization

As illustrated in Fig. 1.3, the ITU's IMT-2030 sixth generation (6G) framework focuses on *enhanced capabilities* (refers to the key performance indicator of peak data rate, latency, connection density, mobility, and reliability) as well as *new capabilities* (refers to key parameters of ubiquitous coverage, sensing, sustainability, resilience, and AI in communications). UAVs, with their recent advancements, can be incorporated into wireless networks to enable these new capabilities. Combined with mMIMO technology and beamforming techniques, mmWave band frequencies can be considered for UAV-assisted communications to support high-throughput and low-latency requirements. While UAVs as BS also offer significant advantages as a flying BS, this book mainly focuses on their applications as relays. Furthermore, with regard to the aforementioned benefits of AI, solutions based on AI can be utilized to solve the complex UAV-assisted mMIMO problems.

A preview of this book is sketched in Fig. 1.4, which highlights and clusters the key subjects covered in various chapters. The remainder of this book is organized as follows.

Chapter 2 presents an overview of the UAVs as relays in wireless communications networks. Then, we investigate various beamforming techniques in mMIMO systems and provide the related literature in UAV-assisted communications. Afterward, we discuss different UAV-assisted relaying schemes in mMIMO systems and review the relevant literature on the subject. Finally, we introduce AI and,

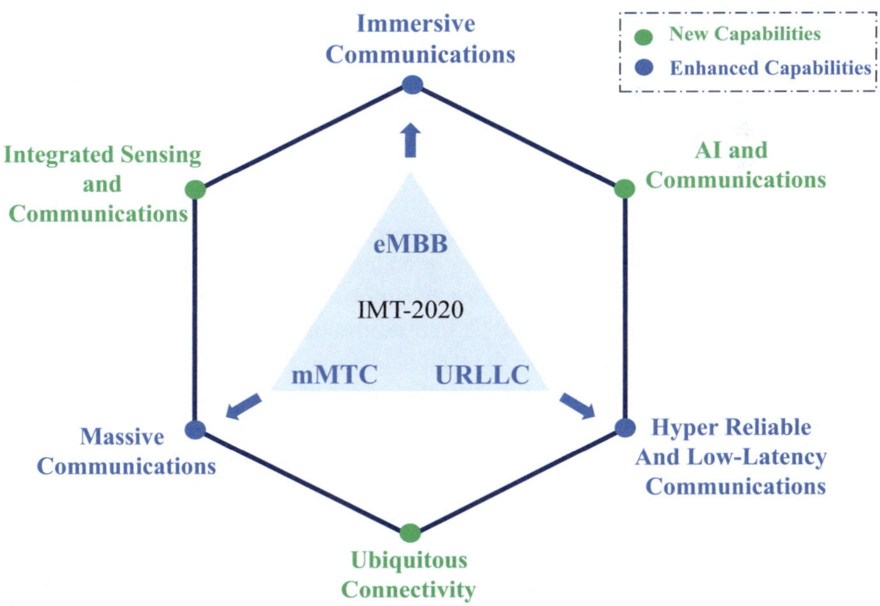

Fig. 1.3 ITU-2030 use cases

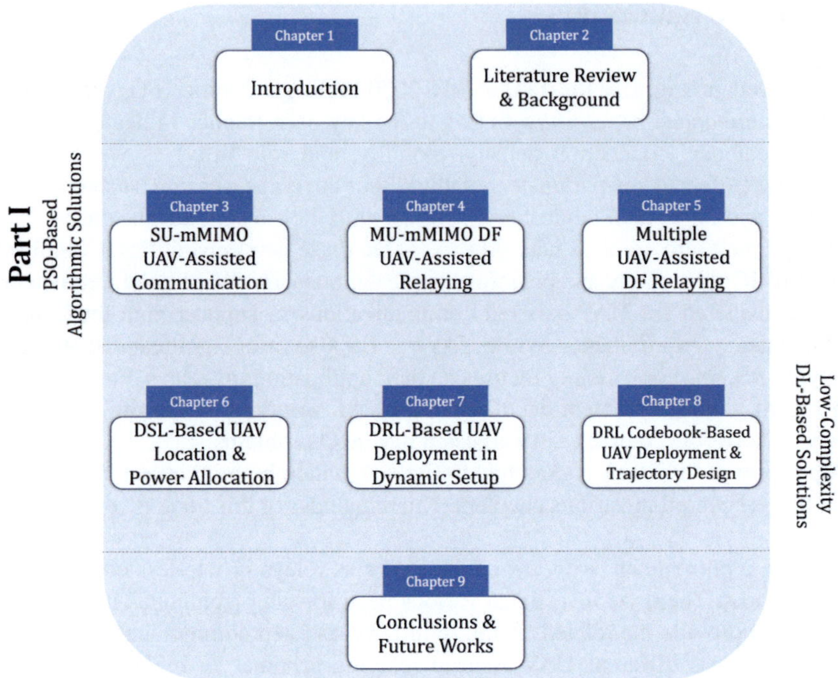

Fig. 1.4 Preview of this book

specifically, NI and deep learning (DL)-based solutions to develop robust and low-complexity solutions for the non-convex optimization problems in UAV-assisted communications.

Chapter 3 investigates various UAV-assisted single-user mMIMO (SU-mMIMO) communications. In particular, we explore complex scenarios where a UAV acts as a relay between a BS and a gateway that, due to possible blockage, the BS does not have direct access to the gateway. We consider HBF setup for the BS and gateway in this chapter. First, we examine a case in which the UAV acts as an amplify-and-forward (AF) relay, while having analog beamforming (ABF) configuration. Then, we propose a joint beamforming and particle swarm optimization (PSO)-based UAV location optimization to maximize the total achievable rate (AR) from the BS to the gateway. The illustrative results show that the proposed joint beamforming and UAV positioning scheme can provide higher spectral and energy efficiencies compared to existing HBF schemes. Then, we use the similar scenario, while this time, the UAV acts as a decode-and-forward (DF) relay and utilizes the HBF configuration. Considering PSO-based UAV deployment, the numerical results indicate that the proposed HBF approach leads to significantly improved energy efficiency when compared to FDBF and can find optimal UAV deployment for enhanced coverage and capacity.

Chapter 4 introduces a UAV-assisted system, where MU-mMIMO systems are considered. We investigate the UAV as the only mean of communication between a BS and remote IoT users, acting as a DF relay. Here, the overall capacity is maximized by optimizing the UAV placement within a deployment area, power allocation (PA) to users jointly with angular-based HBF solutions for both BS and UAV. We introduce various PSO-based algorithms for the UAV deployment and PA optimization. Specifically, jointly with HBF design, we develop three PSO-based solutions: (1) joint HBF, PSO-based PA and fixed UAV location (J-HBF-PSOPA-FL), (2) joint HBF, PSO-based UAV location and equal PA (J-HBF-PSOL-EQPA), and (3) joint HBF, PSO-based UAV location and PA (J-HBF-PSOLPA). Second, we introduce buffer-aided transmission to enhance the capacity. The illustrative results show that the J-HBF-PSOLPA attains the best sum rate capacity compared to other algorithms, where using buffer-aided transmission can further improve the sum rate performance.

Chapter 5 studies the multiple UAVs-assisted networks to support a large number of users, enhancing both coverage and capacity, while the proposed HBF techniques in Chap. 4 follow the use of single UAV. We consider multiple UAVs operating as DF relays, which connect the BS to a large number of IoT devices. We utilize structured sequential optimization to address the multifaceted optimization problem by splitting it into two subproblems. First, K-means-based user clustering is used for UAV-users association. Then, the location of each UAV is optimized jointly with PA using swarm intelligence. The RF stages are designed based on the slow time-varying angular information, while BB stages are designed utilizing the reduced-dimension effective channel matrices. Illustrative results show that multiple UAV-assisted cooperative relaying systems outperform a single UAV system in practical user distributions.

In Part II of this book, we focus on developing low-complexity DL-based solutions for real-time UAV-assisted MU-mMIMO communications problems. In Chap. 6, we aim to reduce the complexity of the proposed J-HBF-PSOLPA algorithm in Chap. 4. For this purpose, we introduce a novel low-complexity joint HBF, DSL-based UAV location and PA (J-HBF-DSLLPA) consisting of two phases: (1) offline deep SL (DSL) using PSO-generated data with J-HBF-PSOLPA and (2) online UAV location and PA prediction. The illustrative results show that the proposed J-HBF-DSLLPA can achieve a sum rate capacity close to J-HBF-PSOLPA, while significantly reducing the runtime.

In Chap. 7, we extend the problem in Chap. 6 by introducing mobility to the IoT users. Likewise, we consider UAV operating as a DF relay between a BS and several IoT users who have no direct link to the BS. All data from the BS reaches the users only through the UAV relay. Considering angular-based HBF for both BS and UAV, the UAV positioning is optimized using a deep RL (DRL) algorithm, namely deep deterministic policy gradient (DDPG) with continuous actions space. The illustrative results show that the proposed DDPG-based UAV deployment (DDPG-UD) can closely approach the AR achieved by PSO-based UAV deployment (PSO-UD), while dramatically reducing the runtime in dynamic environments.

In Chap. 8, we propose a DDPG codebook-based UAV deployment (DDPG-C-UD) for the UAV location optimization and trajectory design, jointly with HBF to further reduce the runtime of the DDPG-UD algorithm in Chap. 7, while considering the same system model. The proposed DDPG-C-UD consists of two distinct phases as: (1) offline agent training using DDPG algorithm and (2) online UAV deployment to maximize the AR. The numerical results demonstrate that the proposed DDPG-C-UD can achieve an AR similar to the DDPG-UD with much lower time complexity. Additionally, the results also indicate that increasing the number of UAVs can remarkably increase the sum rate capacity and improve the coverage.

Finally, Chap. 9 provides the concluding remarks, summarizes the key findings of this book, and discusses the possible future research directions.

References

1. Ericsson, *5G Systems—Enabling the Transformation of Industry and Society*, White Paper, 2017
2. 5G Americas, *LTE to 5G: Cellular and Broadband Innovation*, White Paper, 2017
3. Ericsson, *5G Advanced: Evolution Towards 6G*, White Paper, 2023
4. I. T. U. (ITU), *IMT Traffic Estimates for the Years 2020 to 2030*, Tech. Rep. ITU-R M.2370-0, 2015
5. *Internet of Things Statistics for 2022—Taking Things Apart* (2022). https://dataprot.net/statistics/iot-statistics/
6. Ericsson, *Ericsson Mobility Report*, Tech. Rep., 2023
7. F. Tariq, M.R.A. Khandaker, K.-K. Wong, M.A. Imran, M. Bennis, M. Debbah, A speculative study on 6G. IEEE Wirel. Commun. **27**(4), 118–125 (2020)
8. International Telecommunication Union Radiocommunication Sector (ITU-R), IMT traffic estimates for the years 2020 to 2030. International Telecommunication Union, Tech. Rep., 2015, Includes global mobile data traffic forecast also reported by Cisco
9. 3GPP TR 38.913, *Study on Scenarios and Requirements for Next Generation Access Technologies (Version 14.2.0)*, 3GPP, Tech. Rep., 2017
10. I. T. U. (ITU), *Minimum Requirements Related to Technical Performance for IMT-2020 Radio Interface(s)*, Tech. Rep. ITU-R M.2410-0, 2017
11. 3GPP TS 38.211, *NR; Physical Channels and Modulation (Version 16.7.0)*, 3GPP, Tech. Rep., 2021
12. I. T. U. (ITU), *Guidelines for Evaluation of Radio Interface Technologies for IMT-2020*, Tech. Rep. ITU-R M.2412-0, 2017.
13. Ericsson, *Ericsson Mobility Report*, Tech. Rep., 2020
14. 5G. Americas, *New Services and Applications with 5G Ultra-Reliable Low Latency Communications*, White Paper, 2018.
15. L. Chettri, R. Bera, A comprehensive survey on Internet of Things (IoT) toward 5G wireless systems. IEEE Internet Things J. **7**(1), 16–32 (2020)
16. L. Bai, L. Zhu, X. Zhang, W. Zhang, Q. Yu, Multi-satellite relay transmission in 5G: concepts, techniques, and challenges. IEEE Netw. **32**(5), 38–44 (2018)
17. M. Mozaffari, W. Saad, M. Bennis, Y.-H. Nam, M. Debbah, A tutorial on UAVs for wireless networks: applications, challenges, and open problems. IEEE Commun. Surv. Tutorials **21**(3), 2334–2360 (2019)
18. Q. Wu, R. Zhang, Common throughput maximization in UAV-enabled OFDMA systems with delay consideration. IEEE Trans. Commun. **66**(12), 6614–6627 (2018)

19. K. Meng, D. Li, X. He, M. Liu, Space pruning based time minimization in delay constrained multi-task UAV-based sensing. IEEE Trans. Veh. Technol. **70**(3), 2836–2849 (2021)
20. J. Lyu, Y. Zeng, R. Zhang, Cyclical multiple access in UAV-aided communications: a throughput-delay tradeoff. IEEE Wireless Commun. Lett. **5**(6), 600–603 (2016)
21. W. Feng, J. Wang, Y. Chen, X. Wang, N. Ge, J. Lu, UAV-aided MIMO communications for 5G Internet of Things. IEEE Internet Things J. **6**(2), 1731–1740 (2018)
22. W. Roh et al., Millimeter-wave beamforming as an enabling technology for 5G cellular communications: theoretical feasibility and prototype results. IEEE Commun. Mag. **52**(2), 106–113 (2014)
23. L. Lu, G.Y. Li, A.L. Swindlehurst, A. Ashikhmin, R. Zhang, An overview of massive MIMO: benefits and challenges. IEEE J. Sel. Topics Signal Process. **8**(5), 742–758 (2014)
24. E.G. Larsson, O. Edfors, F. Tufvesson, T.L. Marzetta, Massive MIMO for next generation wireless systems. IEEE Commun. Mag. **52**(2), 186–195 (2014)
25. E. Bjornson, L. Van der Perre, S. Buzzi, E.G. Larsson, Massive MIMO in Sub-6 GHz and mmWave: physical, practical, and use-case differences. IEEE Wireless Commun. **26**(2), 100–108 (2019)
26. J. Lee, J.-K. Han, J. Zhang, MIMO technologies in 3GPP LTE and LTE-advanced. EURASIP J. Wirel. Commun. Netw. **2009**, 1–10 (2009)
27. T.L. Marzetta, Noncooperative cellular wireless with unlimited numbers of base station antennas. IEEE Trans. Wireless Commun. **9**(11), 3590–3600 (2010)
28. A.F. Molisch et al., Hybrid beamforming for massive MIMO: a survey. IEEE Commun. Mag. **55**(9), 134–141 (2017)
29. X. Zhang, J. Li et al., Vehicle-to-everything communication in intelligent connected vehicles: a survey and taxonomy. Autom. Innov. **8**, 13–45 (2025)
30. I. Rojek, P. Kotlarz, J. Dorożyński, D. Miko lajewski, Sixth-generation (6G) networks for improved M2M communication in Industry 4.0. Electronics **13**(8), 1832 (2024)
31. C. Zhang, Y.-L. Ueng, C. Studer, A. Burg, Artificial intelligence for 5G and beyond 5G: implementations, algorithms, and optimizations. IEEE J. Emer. Sel. Topics Circuits Syst. **10**(2), 149–163 (2020)
32. X. Wang, X. Li, V.C.M. Leung, Artificial intelligence-based techniques for emerging heterogeneous network: state of the arts, opportunities, and challenges. IEEE Access **3**, 1379–1391 (2015)
33. C.-X. Wang, M.D. Renzo, S. Stanczak, S. Wang, E.G. Larsson, Artificial intelligence enabled wireless networking for 5G and beyond: recent advances and future challenges. IEEE Wirel. Commun. **27**(1), 16–23 (2020)

Chapter 2
UAVs in the Next Generation Wireless Networks: Background and Literature Review

2.1 An Overview of UAVs as Relay in Wireless Communications Networks

UAVs, commonly referred to as drones, have been a topic of focused research over the past few years [1–4], due to their autonomy, flexibility, and a wide range of uses. Specifically, UAVs have been recognized as key tools for numerous applications, including military operations, surveillance and monitoring, telecommunications, medical supply delivery, and rescue missions [5–9]. These recent advances in drone technology make it possible to widely deploy UAVs, such as drones, small aircraft, balloons, and airships for wireless communications purposes [10–12]. UAVs can be integrated into wireless communications networks, either as flying BSs or as mobile relays. While UAVs offer significant advantages in both roles, this section focuses specifically on their application as relays.

UAVs as relays provide several advantages over conventional static relay systems. Specifically, mobile, on-demand relay systems are well-suited for unexpected or temporary events, such as emergencies or network offloading, due to their capacity for rapid and cost-effective deployment [13]. Furthermore, UAVs' ability to dynamically change their location allows them to optimize deployment, creating ideal communication conditions that are particularly advantageous for delay-tolerant applications like periodic sensing and large data transfers [12, 14, 15]. In addition to these benefits, UAVs' ability to fly in high altitudes enables them to establish a dominant LoS communications channel with ground users, avoiding signal interference from obstacles and making them particularly appealing for dynamic communication systems [16].

UAV has a great capability in reaching inaccessible locations and establishing a LoS link with Ground Users, thereby supporting high-rate communications. However, as data rate demands continue to grow, the overcrowded sub-6 GHz frequency bands are increasingly unable to keep pace. Millimeter-wave communication, with its ample spectrum resources, offers a promising alternative, particularly

T. Le-Ngoc et al., *AI-Enabled UAV-Assisted Massive MIMO*, Wireless Networks,
https://doi.org/10.1007/978-3-032-07872-8_2

suited for UAV applications requiring high throughput and low latency. Despite its potential, mmWave communications face considerable challenges, including high propagation losses from free-space path loss, atmospheric and molecular absorption, and rain attenuation. To address these challenges, several beamforming methods using mMIMO technology have been introduced. In the next section, we take a closer look at different beamforming designs for mMIMO systems to choose the best architecture for UAV-assisted communications.

2.2 Beamforming Techniques in mMIMO Systems

Beamforming is a fundamental technique in mMIMO systems, enhancing communication performance by directing signal energy toward desired directions. This targeted focus is achieved through the use of large antenna arrays that adjust the phase and amplitude of signals. By optimizing the beamforming weights, mMIMO systems can achieve a high SINR, which is crucial for reliable communications. As the number of antennas increases, the SINR also increases, which is referred to as beamforming gain [17]. This gain allows mMIMO systems to direct the signal energy toward desired regions, reducing interference and improving signal quality. This high beamforming gain is particularly effective in tackling the significant path loss discussed earlier in mmWave communications. Moreover, the spacing between antenna elements also plays a critical role in beamforming performance. Typically, antenna separation is maintained at half the wavelength of the carrier frequency to avoid grating lobes and ensure optimal spatial resolution. This spacing is crucial, as it impacts the overall beampattern, sidelobe levels, and beamforming gain. As shown in Fig. 2.1, the three main architectures for beamforming in mMIMO systems are:

- Single-stage FDBF
- Single-stage fully ABF
- Two-stage HBF

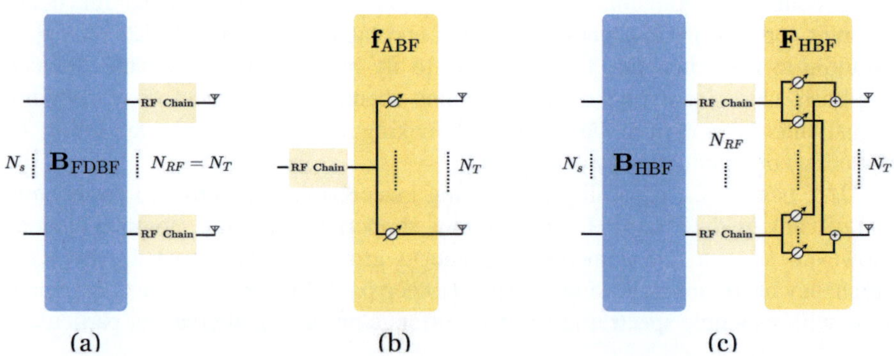

Fig. 2.1 Beamforming architectures for mMIMO systems: (**a**) FDBF, (**b**) ABF, and (**c**) HBF

Table 2.1 Advantages and limitations of beamforming techniques in mMIMO systems

Aspect	FDBF	ABF	HBF
Advantages	Offers precise control of amplitude and phase per antenna, enabling optimal beamforming and high spectral efficiency	Low cost and power efficient as it uses a single RF chain with PSs, making it suitable for large arrays	Balances power consumption and performance; enables multi-stream transmission with fewer RF chains than FDBF
Limitations	High hardware cost and power consumption due to one RF chain per antenna element	Limited flexibility; only controls phase, so performance is lower in MU scenarios	Increased complexity compared to ABF; requires optimization of both RF and BB stages

These architectures have different advantages and limitations. Understanding the strengths and weaknesses of each can help in selecting the appropriate beamforming strategy for the UAV-assisted applications. A brief summary of the advantages and limitations of each beamforming solution is presented in Table 2.1.

2.2.1 FDBF

In traditional MIMO systems, typically featuring a limited number of antennas, a single-stage FDBF architecture is commonly employed. This approach performs beamforming directly at the BB level using digital signal processing techniques [18–20]. FDBF offers significant flexibility and high degrees of freedom (DoF), making it possible to implement efficient beamforming algorithms. As illustrated in Fig. 2.1a, the FDBF architecture supports N_s data streams using N_T antennas, where $N_s \leq N_T$. The digital BB stage for downlink or the digital BB combiner for uplink is represented by the matrix $\mathbf{B}_{\text{FDBF}} \in \mathbb{C}^{N_T \times N_s}$. This matrix contains the beamforming weights, which can be adjusted in both magnitude and phase using digital processing techniques. FDBF facilitates the transmission of multiple data streams simultaneously by effectively mitigating interference, such as inter-symbol interference (ISI) in SU-mMIMO and inter-user interference (IUI) in MU-mMIMO.

Although FDBF can achieve optimal spectral efficiency in mMIMO systems, its design is impractical for UAV-assisted applications. This is because each antenna element in the FDBF architecture requires its own dedicated RF chain, which is both power intensive and costly. Consequently, for a UAV equipped with N_T antennas, there are N_T RF chains, leading to significantly increased hardware costs and power consumption, making FDBF infeasible for UAVs with limited power resources [21]. Furthermore, as the number of antennas increases, so does the computational complexity associated with beamforming optimization, due to the expanding matrix dimensions of $\mathbf{B}_{\text{FDBF}} \in \mathbb{C}^{N_T \times N_s}$. Given these challenges, it is important to explore alternative beamforming architectures that lower hardware and computational demands, while still achieving near-optimal spectral efficiency for UAV applications.

2.2.2 ABF

Figure 2.1b depicts the ABF architecture, where a single RF chain supports N_T antennas [22]. In this setup, N_T low-cost PSs connect the RF chain to the antennas. These PSs consume significantly less power compared to RF chains, typically around 1 mW per PS versus 250 mW per RF chain. This substantial reduction in power consumption and hardware costs makes ABF particularly beneficial for UAV-enabled communications. The use of a single RF chain in the ABF architecture remains constant regardless of the number of antennas, significantly reducing hardware complexity and costs. The analog RF beamforming vector is represented as $\mathbf{f}_{\text{ABF}} \in \mathbb{C}^{N_T}$. Compared to the digital BB beamforming matrix $\mathbf{B}_{\text{FDBF}} \in \mathbb{C}^{N_T \times N_s}$, which has $N_T \times N_s$ elements, $\mathbf{f}_{\text{ABF}}$ with N_T elements exhibits much lower computational complexity [23].

Despite its advantages, the ABF architecture has some significant limitations. The analog RF beamforming vector $\mathbf{f}_{\text{ABF}}$ typically only applies phase shifts to the transmitted and received signals, adhering to the constant modulus (CM) constraint. This limitation complicates beamforming optimization and reduces the DoF compared to FDBF, which can adjust both the magnitude and phase of signals. As a result, the ABF architecture generally experiences performance degradation compared to FDBF. Moreover, ABF supports only a single data stream, while FDBF, as shown in Fig. 2.1a, can handle multiple data streams, which is beneficial especially for MU-mMIMO scenarios. This limitation reduces spectral efficiency, which is one of the key benefits of mMIMO systems.

Thus, there is a trade-off between digital and analog beamforming architectures. FDBF offers higher spectral efficiency by supporting multiple data streams and providing greater flexibility in signal processing. On the other hand, ABF improves hardware and computational efficiency by reducing power consumption and complexity. In this case, we explore the third beamforming architecture that balances the advantages and drawbacks of both FDBF and ABF.

2.2.3 HBF

The two-stage HBF architecture has emerged as a promising solution for UAV-assisted mMIMO systems, effectively addressing the trade-offs between single-stage FDBF and ABF architectures [24, 25]. As illustrated in Fig. 2.1c, HBF divides the beamforming process into two stages: an analog transmit (or receive) RF beamformer $\mathbf{F}_{\text{HBF}} \in \mathbb{C}^{N_T \times N_{\text{RF}}}$ and a digital BB precoder (or combiner) $\mathbf{B}_{\text{HBF}} \in \mathbb{C}^{N_{\text{RF}} \times N_s}$ for transmission (or reception). This two-stage approach leverages a reduced number of RF chains, N_{RF}, where $N_s \leq N_{\text{RF}} \leq N_T$, to minimize hardware cost and complexity in UAV communications with limited power resources. Additionally, the HBF architecture supports multiple data streams, making it a versatile solution.

In the RF stage, each RF chain connects to every antenna element through PSs, constructing the analog RF beamformer $\mathbf{F}_{\text{HBF}} \in \mathbb{C}^{N_T \times N_{\text{RF}}}$ with $N_T \times N_{\text{RF}}$ PSs. Subsequently, the digital BB precoder/combiner $\mathbf{B}_{\text{HBF}} \in \mathbb{C}^{N_{\text{RF}} \times N_s}$ is designed in the BB stage. By employing a reduced number of RF chains, the two-stage HBF architecture significantly cuts down on hardware costs and power consumption. Moreover, HBF techniques optimize both RF and BB stages to enhance system capacity. Numerical studies indicate that HBF can closely match the spectral efficiency of FDBF, while significantly outperforming ABF. Thus, HBF achieves high spectral, energy, and hardware efficiency for UAV-enabled mMIMO systems, offering more DoFs for low-complexity HBF techniques.

Most existing research on FDBF and HBF architectures assumes the availability of full-size instantaneous CSI [26–32]. However, as antenna arrays scale up, the channel dimension and pilot transmission length increase, resulting in substantial channel estimation overhead. The HBF architecture can mitigate this overhead by developing the RF stage using slow time-varying CSI (e.g., channel covariance matrix or angular information), and subsequently designing the BB stage with reduced-size effective instantaneous CSI. This capability further enhances the practicality of HBF by reducing the channel estimation overhead. HBF architectures are categorized into fully connected (FC) and sub-connected (SC) configurations. In FC HBF architecture, each RF chain connects to all antenna elements, which necessitates $N_T \times N_{\text{RF}}$ PSs. Conversely, SC HBF architecture reduces the number of PSs to N_T by connecting each RF chain to a subset of antennas. It is worthwhile to note that we focus on the FC HBF technique for UAV-assisted mMIMO systems in the context of this book. Table 2.2 provides a comparison between ABF, FDBF, and HBF, which shows that advanced HBF techniques can achieve high spectral efficiency with low hardware cost/complexity, low CSI overhead size for practical UAV-assisted mMIMO systems.

2.2.4 HBF in UAV-Assisted Communications

Due to the aforementioned advantages, recent research has shown growing interest in adopting HBF architecture for UAV-assisted communications [33–39]. Specifically, in [33], an optimization problem for UAV location, user clustering, and HBF design is presented to maximize capacity under a minimum rate constraint for each user. This is achieved using an alternating optimization, successive convex optimization, and combinatorial optimization scheme. Similarly, [34] proposes an optimal PA and HBF design for UAV-assisted mMIMO communications and derives a closed-form expression of rate for LoS channels. The authors show that the total transmit power should be inversely proportional to the number of

Table 2.2 Comparison of fully digital, analog, and hybrid beamforming in UAV-enabled mMIMO systems

Aspect	FDBF	ABF	HBF
Hardware complexity	High: One RF chain per antenna	Low: Single RF chain, PSs	Medium: Few RF chains, PSs
Power consumption	High	Low	Medium
Computational complexity	High	Low	Medium
Flexibility	High: Change magnitude and phase	Low: Change phase only	Medium: Combines FDBF and ABF
Spectral efficiency	High: Optimal	Low: Limited by single stream	High: Near-optimal, multiple streams
Cost	High	Low	Medium
Implementation feasibility	Difficult for large arrays	Easy, cost-effective	Feasible with trade-offs
CSI overhead	High	Medium: Slow time-varying CSI	Medium: Effective CSI

UAV antennas to meet each user's rate requirement. In [35], the energy efficiency of mmWave-enabled non-orthogonal multiple access (NOMA) is maximized by jointly optimizing UAV location, HBF, and PA through decomposition of the non-convex problem into several subproblems. In [36], the joint optimization of UAV's three-dimensional (3-D) placement, HBF, and resource allocation is addressed to maximize the sum rate, while satisfying the constraints of energy harvesting and coverage using a polyhedral annexation method and deep deterministic policy gradient algorithm. Reference [37] considers UAV position and HBF optimization by employing UAVs as flying BSs, where the non-convex problem is solved using a two-stage optimization strategy. First, the UAV is deployed based on an ideal beampattern assumption, and then the optimal location is used for HBF design by considering the UAV jitter. In [38], block coordinate descent (BCD)-based algorithmic solutions are proposed for joint PA, ABF, and UAV positioning to maximize energy efficiency, while satisfying constraints such as maximum transmitting power, deployment span, and minimum data rate from the ground users. Similarly, in [39], the joint optimization problem of UAV 3-D placement, beampattern, and transmit power is considered. The non-convex problem is solved using a multi-objective evolutionary algorithm based on decomposition (MOEA/D)-based algorithm.

After selecting the appropriate beamforming architecture for UAV-assisted mMIMO systems, we will examine various UAV relay mMIMO techniques in the context of HBF in the next section.

2.3 UAV-Assisted mMIMO Relaying

Relaying techniques in MIMO communications, particularly in the context of HBF and for UAV-assisted communications, play a vital role in enhancing coverage, reliability, and overall system performance. This section explores the two primary relaying strategies, AF and DF, in the context of HBF [40–43].

2.3.1 *AF UAV Relaying*

In AF relaying, the UAV relay amplifies the received signal and forwards it to the destination. Considering an HBF architecture, the UAV utilizes both analog and digital beamforming stages to process the signal efficiently. For a mMIMO communications system with N_T antennas at the transmitter, N_R antennas at the UAV, and N_D antennas at the destination, the received signal at the UAV relay is expressed as

$$\mathbf{y}_R = \mathbf{H}_{SR}\mathbf{x}_S + \mathbf{n}_R, \tag{2.1}$$

where $\mathbf{H}_{SR} \in \mathbb{C}^{N_R \times N_T}$ is the channel matrix between the source and the UAV, $\mathbf{x}_S \in \mathbb{C}^{N_T \times 1}$ is the transmitted signal from the source, and $\mathbf{n}_R \in \mathbb{C}^{N_R \times 1}$ is the noise vector at the UAV relay. Considering an HBF architecture at the UAV, the received signal is amplified and forwarded to the destination. Then, the signal transmitted from the UAV is given by

$$\mathbf{x}_R = \mathbf{F}\mathbf{B}\mathbf{y}_R, \tag{2.2}$$

where $\mathbf{F} \in \mathbb{C}^{N_R \times N_{RF}}$ is the analog beamforming matrix, and $\mathbf{B} \in \mathbb{C}^{N_{RF} \times N_s}$ is the digital beamforming matrix. The received signal at the destination is

$$\mathbf{y}_D = \mathbf{H}_{RD}\mathbf{x}_R + \mathbf{n}_D = \mathbf{H}_{RD}\mathbf{F}\mathbf{B}\mathbf{y}_R + \mathbf{n}_D. \tag{2.3}$$

Substituting $\mathbf{y}_R$ into the above equation, we get

$$\mathbf{y}_D = \mathbf{H}_{RD}\mathbf{F}\mathbf{B}(\mathbf{H}_{SR}\mathbf{x}_S + \mathbf{n}_R) + \mathbf{n}_D. \tag{2.4}$$

The design of HBF stages **F** and **B** is crucial to balance the trade-off between signal amplification and noise control. The design of HBF matrices can maximize the received signal power, while minimizing the amplified noise.

2.3.2 DF UAV Relaying

In DF relaying, the UAV relay decodes the received signal, re-encodes it, and then forwards it to the destination. The HBF technique can be used at both the UAV and the destination to process the signal efficiently. The received signal at the UAV in DF relaying is the same as in AF relaying:

$$\mathbf{y}_R = \mathbf{H}_{SR}\mathbf{x}_S + \mathbf{n}_R. \tag{2.5}$$

The UAV decodes the received signal to obtain $\hat{\mathbf{x}}_S$, which is an estimate of the transmitted signal $\mathbf{x}_S$. The UAV then re-encodes and forwards $\hat{\mathbf{x}}_S$ to the destination using HBF architecture. The transmitted signal from the UAV is

$$\mathbf{x}_R = \mathbf{FB}\hat{\mathbf{x}}_S. \tag{2.6}$$

Then, the received signal at the destination is

$$\mathbf{y}_D = \mathbf{H}_{RD}\mathbf{x}_R + \mathbf{n}_D = \mathbf{H}_{RD}\mathbf{FB}\hat{\mathbf{x}}_S + \mathbf{n}_D. \tag{2.7}$$

In practical scenarios, decoding errors at the UAV can degrade performance. The joint design of HBF stages at BS, UAV, and at destination can maximize the sum rate capacity by improving SINR.

The choice between AF and DF UAV relaying depends on the specific requirements of the communication system. AF relaying with HBF is simpler and introduces less delay, but can amplify noise. DF relaying with HBF provides better performance in terms of noise mitigation but requires more complex processing and introduces decoding delays. A brief comparison of the DF versus AF relaying scheme is presented in Table 2.3.

2.3.3 UAV Relaying in the Literature

The deployment of UAVs as relays—not limited to mMIMO systems—has attracted significant research interest in recent years, aiming to design UAV-assisted systems that either maximize throughput or minimize transmit power [44–50]. In particular,

Table 2.3 Comparison of relaying protocols in mMIMO systems

Criterion	AF relaying	DF relaying
Complexity	Moderate	High
Noise amplification	Yes	No
Processing delay	Low	High
Implementation	Moderate	Complex
Reliability	Moderate	High

the authors in [44] propose a UAV relay networks optimization method that minimizes outage probability through joint trajectory design and power control. Their proposed solution reduces outage probability in comparison to a fixed power relay and a circular UAV trajectory. In [45], the optimal altitude for a UAV relay in communications systems is analyzed, and the authors demonstrate that different reliability measures have slightly different optimum altitudes. Additionally, they show that the DF relay is superior to AF relay. In [46], a single UAV is used as a mobile relay for wireless communications networks in harsh environments. The source/relay PA and UAV trajectory are optimized to maximize the end-to-end throughput. Similarly, in [47], a UAV serves as a mobile relay in cases of disrupted communications lines during emergencies, and a variable-rate-relaying strategy is proposed to reduce the risk of outages and improve data rate. The authors in [48] propose a MU UAV-relaying system that maximizes the sum rate of ground users, while satisfying cellular user rate requirements through the optimization of UAV placement and transmit PA. The non-convex problem is solved using BCD and successive convex optimization. In [49], an interference coordination approach using UAV relaying is proposed to enhance data transmission reliability and alleviate mutual interference among ground users by jointly optimizing throughput and UAV energy consumption using the BCD method. A UAV relay is used in an uplink MIMO cognitive radio system in [50] to enable communications between a primary user and a secondary user. The optimal PA is derived to maximize the AR of the secondary user, while respecting power budget, interference, and relay power constraints.

2.4 UAV-Assisted Relaying in Multiuser Massive MIMO Configuration

One of the key features of mMIMO is to support multiuser communications (i.e., MU-mMIMO), where multiple single-antenna users ($K \leq N_T$) are served simultaneously by employing advanced beamforming techniques. Both downlink and uplink transmissions in MU-mMIMO systems are facilitated by designing appropriate beamforming techniques to enhance the overall sum rate capacity, where here we only focus on the downlink transmission. The primary goals of beamformer design in MU-mMIMO systems are to maximize the desired signal for each user, while effectively mitigating multiuser interference (MUI). MU-mMIMO systems offer significant advantages over SU-mMIMO systems. Firstly, the channel conditions improve due to the spatial diversity provided by users positioned at different locations, leading to a higher channel rank. Secondly, MU-mMIMO systems serve users with fewer antennas, which is more practical for wireless communications systems constrained by hardware limitations and cost considerations.

Let us focus only on the UAV-users communications as the BS-UAV communications is a SU-mMIMO scenario. In the downlink transmission of the FDBF architecture, depicted in Fig. 2.1a, the signal transmitted from the UAV after precoding is represented as $\mathbf{s}_D = \mathbf{B}_D\mathbf{d}_D \in \mathbb{C}^{N_T}$. Here, $\mathbf{B}_D = [\mathbf{b}_{D,1}, \mathbf{b}_{D,2}, \ldots, \mathbf{b}_{D,K}] \in \mathbb{C}^{N_T \times K}$ denotes the digital baseband beamforming matrix, and $\mathbf{d}_D = [d_{D,1}, d_{D,2}, \ldots, d_{D,K}]^T \in \mathbb{C}^K$ is the vector containing the downlink data signals. The received signal at the user side is then expressed as

$$\mathbf{r}_D = \mathbf{H}_D\mathbf{s}_D + \mathbf{n}_D = \mathbf{H}_D\mathbf{B}_D\mathbf{d}_D + \mathbf{n}_D = \begin{bmatrix} r_{D,1} \\ r_{D,2} \\ \vdots \\ r_{D,K} \end{bmatrix} = \begin{bmatrix} \mathbf{h}_{D,1}^T\mathbf{B}_D\mathbf{d}_D \\ \mathbf{h}_{D,2}^T\mathbf{B}_D\mathbf{d}_D \\ \vdots \\ \mathbf{h}_{D,K}^T\mathbf{B}_D\mathbf{d}_D \end{bmatrix} + \mathbf{n}_D, \tag{2.8}$$

where $\mathbf{H}_D = [\mathbf{h}_{D,1}, \mathbf{h}_{D,2}, \ldots, \mathbf{h}_{D,K}]^T \in \mathbb{C}^{K \times N_T}$ represents the downlink channel matrix, $\mathbf{h}_{D,k} \in \mathbb{C}^M$ corresponds to the channel vector for the kth user, and $\mathbf{n}_D \sim \mathcal{CN}(0, \sigma_n^2\mathbf{I}_K)$ is the complex Gaussian noise vector. The primary goal of the digital baseband beamforming stage $\mathbf{B}_D$ is to mitigate MUI, while maximizing the desired signal power. Common linear beamforming methods for designing the digital BB precoder include: (i) matched filter (MF), (ii) zero-forcing (ZF), and (iii) regularized zero-forcing (RZF). The general form of the digital BB precoder is given by

$$\mathbf{B}_D = \alpha_D\mathbf{X}_D^{-1}\mathbf{H}_D^H \in \mathbb{C}^{N_T \times K}, \tag{2.9}$$

where α_D is a normalization factor ensuring that the transmit power constraint P_T is met, and $\mathbf{X}_D \in \mathbb{C}^{N_T \times N_T}$ varies based on the technique: $\mathbf{X}_D = \mathbf{I}_{N_T}$ for MF, $\mathbf{X}_D = \mathbf{H}_D^H\mathbf{H}_D$ for ZF, and $\mathbf{X}_D = \mathbf{H}_D^H\mathbf{H}_D + \frac{\sigma_n^2}{P_T/K}\mathbf{I}_{N_T}$ for RZF.

Beyond the single-stage conventional FDBF techniques, the two-stage HBF architecture discussed in Sect. 2.2.3 can also be utilized in MU-mMIMO systems to enhance spectral efficiency, while reducing hardware complexity. However, the single-stage ABF relies solely on phase control and can generate only one spatial stream. This limitation leads to much higher MUI compared with HBF and FDBF, making ABF unsuitable for MU-mMIMO scenarios.

2.5 AI/ML Applications in UAV-Assisted Communications

In recent years, the use of AI in wireless communications has attracted a lot of interest, thanks to its ability to provide efficient and reliable solutions for challenging optimization problems [51]. The superiority of AI-based solutions has been verified in certain applications, such as UAV-assisted wireless communications [52–54]. As depicted in Fig. 2.2, AI encompasses various branches, including nature-inspired (NI) algorithms and several types of ML approaches. These ML approaches include

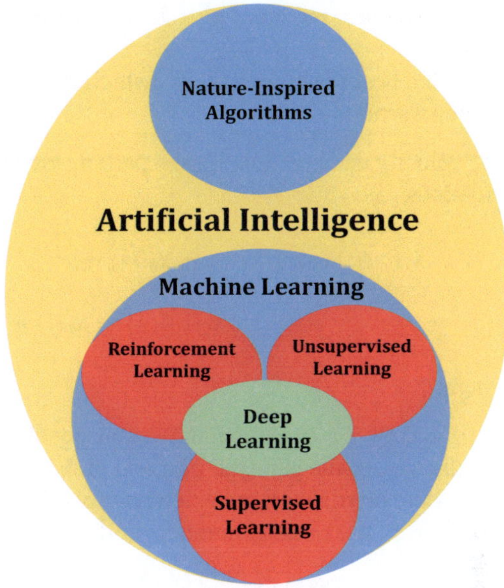

Fig. 2.2 Relationship of AI and its branches

supervised learning (SL), unsupervised learning (UL), reinforcement learning (RL), and DL. In this section, we first present PSO as an NI algorithm, and then, we discuss SL, RL, and DL and their potential applications in UAV-assisted communications.

2.5.1 NI Intelligent Algorithms

NI algorithms have gained popularity due to their effectiveness in solving complex optimization problems. Unlike traditional deterministic methods, these algorithms employ multiple search agents, which are initially randomly placed in the optimization space and move toward the global optimum through iterative communications, combining deterministic and stochastic movements. Some of the key NI algorithms are: PSO, genetic algorithm (GA), ant colony optimization (ACO), etc.

Among different NI algorithms, PSO is inspired by the social behavior of animals such as birds flocking or fish schooling [55]. It is used to solve optimization problems by iteratively improving candidate solutions with regard to a given measure of quality. The basic idea of PSO involves a swarm of particles moving through the search space to find the optimal solution. Each particle represents a potential solution and has the following attributes:

- Position: $\mathbf{x}_i \in \mathbb{R}^n$ for particle i
- Velocity: $\mathbf{v}_i \in \mathbb{R}^n$ for particle i

- Personal best position: $\mathbf{p}_i \in \mathbb{R}^n$, which is the best position found by particle i so far
- Global best position: $\mathbf{g} \in \mathbb{R}^n$, which is the best position found by any particle in the swarm

The velocity and position of each particle are updated iteratively using the following equations:

$$\mathbf{v}_i(t+1) = \gamma_1 z_1(\mathbf{p}_i - \mathbf{x}_i(t)) + \gamma_2 z_2(\mathbf{g} - \mathbf{x}_i(t)) + \gamma_3 \mathbf{v}_i(t), \tag{2.10}$$

$$\mathbf{x}_i(t+1) = \mathbf{x}_i(t) + \mathbf{v}_i(t+1), \tag{2.11}$$

where:

- γ_1 and γ_2 are cognitive and social coefficients, respectively, which control the influence of the personal best and global best positions.
- ω is the inertia weight, which controls the impact of the previous velocity.
- z_1 and z_2 are random numbers uniformly distributed in [0, 1].

The iterative process of PSO can be summarized as follows:

1. Initialize the positions $\mathbf{x}_i(0)$ and velocities $\mathbf{v}_i(0)$ of all particles randomly.
2. Evaluate the fitness of each particle's position.
3. Update each particle's personal best position $\mathbf{p}_i$ if the current position is better.
4. Update the global best position $\mathbf{g}$ if any particle's personal best position is better than the current global best.
5. Update the velocities and positions of all particles using Eqs. (2.10) and (2.11).
6. Repeat steps 2–5 until a stopping criterion is met (e.g., a maximum number of iterations or a satisfactory fitness level).

2.5.2 SL

SL is a type of ML where an algorithm is trained on labeled data. In this approach, the dataset consists of input-output pairs, with each input (often called features) mapped to a specific, correct output (label or target). The algorithm learns to map inputs to the correct outputs by finding patterns in the labeled data and minimizing the difference between its predictions and the actual labels [52].

SL is extensively used in UAV-assisted communications for its ability to simplify complex computations, significantly reducing runtime [56–59]. Specifically, the authors in [56] investigate an optimal placement of UAVs acting as aerial BSs by building a structured radio map. The authors propose a joint clustering and regression problem using a maximum likelihood approach that is formulated based on the K-segment ray-tracing model. Similarly, in [57], the authors study the optimal UAV deployment. SL techniques are used to predict download traffic using weighted expectation maximization. This ML technique has been compared in terms of performance to a baseline expectation maximization algorithm and the K-means algorithm. The authors in [58] investigate generating waypoints for the UAV flight path based on previous flight records. In [59], the authors improve the efficiency between a UAV and BS by predicting the location of the UAV given its past locations.

2.5.3 RL

RL is a subfield of ML, where an intelligent agent takes action sequentially to maximize a numerical reward [60]. The agent must learn to make decisions that maximize rewards, despite having no prior knowledge of the environment or the potential outcomes of its actions. This involves exploring different strategies, observing the results, and continuously adapting its actions based on past experiences to improve future performance.

RL differs fundamentally from SL. In SL, the model learns from a provided set of labeled data, identifying relationships between inputs and outputs to make predictions on new, unseen inputs. RL, on the other hand, is designed for interactive environments where an agent learns optimal behavior through experience, aiming to maximize a long-term reward. RL is particularly valuable in scenarios that require optimizing a sequence of decisions or where labeled outputs are not available.

The RL involves a sequential decision-making process where the goal is to maximize total future reward. As illustrated in Fig. 2.3, at each time step t, the agent receives a representation of the environment, $\mathbf{s}_t$, called the state vector. Based on the state $\mathbf{s}_t$, the agent takes the action vector $\mathbf{a}_t$, resulting in a numerical reward r_{t+1} receives the feedback reward r_{t+1}, and the environment transitions into a new state $\mathbf{s}_{t+1}$. The action taken at a specific time instance can have long-term consequences.

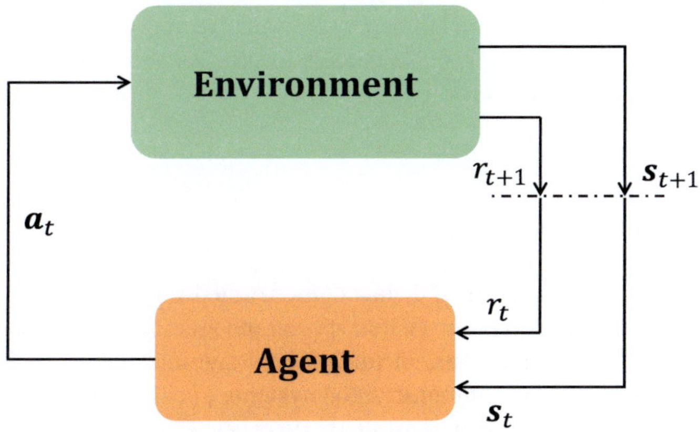

Fig. 2.3 An agent-environment interaction

In some cases, it might be better to sacrifice the immediate reward for a greater cumulative reward in the future. For example, in a game of chess, capturing an opponent's piece might not be a good idea if it results in a state that increases the likelihood of loss.

RL has been extensively used in different fields of robotics and autonomous vehicles [61–64]. The significant success of RL in these areas has inspired researchers to investigate its potential applications within UAV-assisted wireless communications [65–69]. In particular, the authors in [65] propose a 3-D UAV trajectory design aimed at enhancing the network's energy efficiency. They achieve this by maximizing an objective function that balances energy consumption and throughput, while ensuring the quality of service (QoS) requirements of ground users are met. The problem is modeled as a multiarmed bandit (MAB) using RL, and they introduce an upper confidence bound (UCB)-based algorithm to solve it. In [66], the authors apply the SARSA (λ) RL method to improve UAV path planning in the investigated environment. Using 3-D modeling for position and trajectory analysis, the method introduces eligibility traces to optimize the SARSA algorithm, resulting in faster learning and smoother state updates. Simulation results show this approach significantly improves the efficiency of UAV path exploration compared to traditional algorithms. Similarly, the authors in [67, 68] propose Q-learning as an RL approach for UAV 3-D path planning to achieve the optimal path. The authors in [69] investigate the self-directed resource allocation in UAV-powered communications networks, formulated as a stochastic game to address dynamic, unpredictable environments. Using an multi-agent RL (MADRL) framework, each UAV independently optimizes long-term rewards via an SARSA-based, agent-agnostic decision process. Simulations show the proposed approach reduces the need for extensive information exchange, enhancing efficiency.

While conventional RL and SL approaches demonstrate acceptable performance in problems where the complexity is manageable, they often fail to achieve satisfactory results when confronted with high-dimensional variables. To address this challenge, DL-based solutions have been proposed for both SL and RL. In the following subsection, we explore DL as a transformative approach that enhances RL and SL capabilities by enabling these models to process and learn from complex, high-dimensional data effectively.

2.5.4 DL

As a cornerstone of AI and ML, DL has demonstrated remarkable success across various fields, including computer vision, speech recognition, and natural language processing [70]. This success has, in turn, driven significant interest in applying DL techniques to wireless communications systems [71–73]. In this domain, DL has proven effective for tasks such as signal detection [71], resource management [72], and channel estimation [73], offering advanced capabilities to enhance system performance and efficiency in complex communication environments.

Fig. 2.4 A single neuron

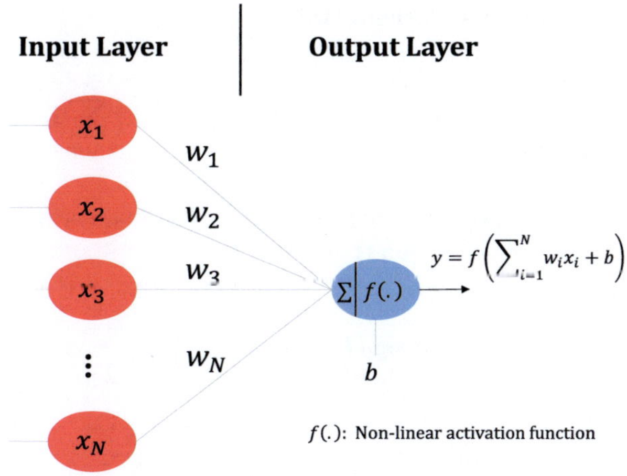

The DL architecture is built upon neural networks and consists of two distinct phases: (1) forward pass and (2) backward pass. The fundamental building block of DL setups is an individual neuron. As it is shown in Fig. 2.4, in the forward pass phase, the neuron takes N inputs $x_1, x_2, x_3, \ldots, x_N$, multiplies each by its respective weight $w_1, w_2, w_3, \ldots, w_N$, sums them with a bias b, and then applies a nonlinear activation function $f(\cdot)$ to the result. The activation function is used to introduce nonlinearity into the model, enabling it to learn and approximate complex patterns and relationships in data beyond simple linear transformations. Consequently, the output of the single-neuron model during the forward pass phase can be expressed as

$$y = f\left(\sum_{i=1}^{N} w_i x_i + b\right). \tag{2.12}$$

Here, common examples of nonlinear activation functions are rectified linear unit (ReLU) $f(x) = \max(0, x)$, leaky ReLU $f(x) = \max(0.1x, x)$, sigmoid $f(x) = \frac{1}{1+e^{-x}}$, and hyperbolic tangent $f(x) = \tanh(x)$ [70].

Conversely, backward pass is used to optimize the weights and biases to achieve an optimal input-output relationship. For this purpose, we minimize the loss function that measures the error between predicted and actual values. Some examples of common loss functions, which will be used in this book, are written as:

- **Mean absolute error (MAE)**:

$$\mathcal{L}_{\text{MAE}} = |\hat{y} - y| = \left|\hat{y} - f\left(\sum_{i=1}^{N} x_i w_i + b\right)\right| \tag{2.13}$$

- **Mean square error (MSE)**:

$$\mathcal{L}_{\mathrm{MSE}} = \left(\hat{y} - y\right)^2 = \left(\hat{y} - f\left(\sum_{i=1}^{N} x_i w_i + b\right)\right)^2 \tag{2.14}$$

where $\hat{y}$ denotes the true value. Then, the gradient with respect to the output neuron for each loss function is given as

$$\frac{\partial \mathcal{L}_{\mathrm{MAE}}}{\partial y} = \operatorname{sign}\left(y - \hat{y}\right) = \operatorname{sgn}\left(f\left(\sum_{i=1}^{N} x_i w_i + b\right) - \hat{y}\right), \tag{2.15}$$

$$\frac{\partial \mathcal{L}_{\mathrm{MSE}}}{\partial y} = 2\left(y - \hat{y}\right) = 2\left(f\left(\sum_{i=1}^{N} x_i w_i + b\right) - \hat{y}\right). \tag{2.16}$$

Subsequently, the gradient of the output with respect to the weight w_i and bias b can be written as

$$\frac{\partial y}{\partial w_i} = f'\left(x\right)|_{x=y} \cdot x_i, \tag{2.17}$$

$$\frac{\partial y}{\partial b} = f'\left(x\right)|_{x=y}, \tag{2.18}$$

where $f'(x)$ is the derivative of the activation function. Next, using the multivariate chain rule of calculus, we obtain the gradient of the loss function with respect to the weights and bias. Thus, for MAE loss function, we will have

$$\frac{\partial \mathcal{L}_{\mathrm{MAE}}}{\partial w_i} = \frac{\partial \mathcal{L}_{\mathrm{MAE}}}{\partial y} \cdot \frac{\partial y}{\partial w_i} = \operatorname{sgn}\left(y - \hat{y}\right) \cdot f'\left(x\right)|_{x=y} \cdot x_i, \tag{2.19}$$

$$\frac{\partial \mathcal{L}_{\mathrm{MAE}}}{\partial b} = \frac{\partial \mathcal{L}_{\mathrm{MAE}}}{\partial y} \cdot \frac{\partial y}{\partial b} = \operatorname{sgn}\left(y - \hat{y}\right) \cdot f'\left(x\right)|_{x=y}. \tag{2.20}$$

Similarly, we can write for MSE loss function as

$$\frac{\partial \mathcal{L}_{\mathrm{MSE}}}{\partial w_i} = \frac{\partial \mathcal{L}_{\mathrm{MSE}}}{\partial y} \cdot \frac{\partial y}{\partial w_i} = 2\left(y - \hat{y}\right) \cdot f'\left(x\right)|_{x=y} \cdot x_i, \tag{2.21}$$

$$\frac{\partial \mathcal{L}_{\mathrm{MSE}}}{\partial b} = \frac{\partial \mathcal{L}_{\mathrm{MSE}}}{\partial y} \cdot \frac{\partial y}{\partial b} = 2\left(y - \hat{y}\right) \cdot f'\left(x\right)|_{x=y}. \tag{2.22}$$

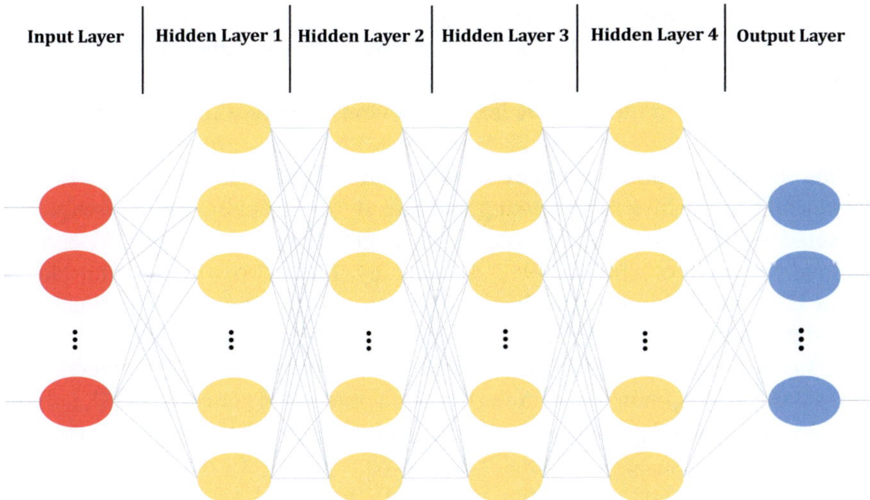

Fig. 2.5 Deep neural network (DNN) architecture

Finally, the gradient descent for weight w_i and bias b updates is formulated as

$$w_i \leftarrow w_i - \alpha \frac{\partial \mathcal{L}_{\mathrm{k}}}{\partial w_i}, \tag{2.23}$$

$$b \leftarrow b - \alpha \frac{\partial \mathcal{L}_{\mathrm{k}}}{\partial b}, \tag{2.24}$$

where $k = \{\mathrm{MAE}, \mathrm{MSE}\}$ and α is the learning rate. Here, these equations are formulated for a single neuron but can be extended to networks with multiple neurons, accommodating more complex architectures that will be used in later chapters of this book.[1]

A neural network is built by interconnecting multiple neurons (outputs), where each input is connected to every neuron, creating a structured, layered architecture. Various neural network architectures have been developed, including deep neural networks (DNNs) [75–77], convolutional neural networks (CNNs) [78–80], and recurrent neural networks (RNNs) [81, 82]. In this book, we focus exclusively on the DNN architecture. An example of DNN architecture is illustrated in Fig. 2.5, where the input and output layers are linked through three hidden layers, with each neuron connected only to neurons in adjacent layers, while neurons within the same layer remain unconnected. Each hidden and output layer can utilize distinct nonlinear activation functions, allowing for flexibility in transforming the data. The depth of

[1] More details (e.g., using minibatch gradient descent or complete mathematical derivations) can be found in [74].

the DNN architecture—defined by the number of hidden layers (HLs)—determines whether it is considered deep or shallow. Additionally, the number of neurons per hidden layer, as well as the overall number of HLs, directly correlates with the network's capacity to model complex input-output relationships in the problem at hand.

This capacity to model complex input-output relationships has made DL architectures highly effective in addressing intricate, dynamic challenges. Consequently, DL-based solutions have found successful applications in various UAV-assisted communications problems [83–91]. Specifically, the authors in [83] introduce a hybrid DSL-based online offloading framework, which predicts the optimal positions of ground vehicles and UAVs in a mobile edge computing platform. In [84], the authors propose a DSL algorithm based on gated recurrent units and autoencoders for UAV trajectory prediction. Similarly, the authors in [85] employ a DSL solution based on GA-based dataset for multi-UAV path planning that can predict the UAV trajectories with reduced computational complexity. In [86], the authors propose a deep Q-network (DQN) to optimize the UAV location based on received signal strength. This approach demonstrates efficient and timely coverage performance. Likewise, the authors in [87] employ DQN to optimize UAV location for AR maximization. For this purpose, they propose a DQN framework to prevent UAV collisions, while keeping them within a set service area. In [88], the authors propose a deep RL (DRL)-based autonomous UAV navigation using a sparse reward function. Using a sparse reward function ensures the agent is rewarded only upon task completion, preventing suboptimal solutions. However, it limits effective learning as informative states are rarely encountered. To address this, the authors leverage a prior policy to assist the agent, demonstrating that their approach outperforms other RL-based algorithms. The authors in [89] introduce UAV-assisted communications system where the UAV's flight time is uncertain and the locations and requirements of the user equipment (UE) are variable. To tackle this challenge, a low-complexity, online solution based on proximal policy optimization 2 (PPO2) is proposed. The solution aims to minimize the UAV's flight time, while simultaneously satisfying other constraints, such as completing communication tasks for the ground UEs. In [90], the authors present a UAV-assisted computation offloading architecture in the THz band, optimizing UAV location, resource allocation, and offloading under QoS and resource constraints. They propose double DQN (DDQN)-based and DDPG-based solutions, both achieving near-optimal results. The authors in [91] study the joint problem of dynamic multi-UAV altitude control and multicell wireless channel access management of IoT devices to enable an energy-sustainable capacity-optimal network. To solve this problem, the authors propose an online model-free constrained DRL (CDRL) algorithm.

2.6 Concluding Remarks

In this chapter, we provided a background and literature survey on UAV-assisted communications in the next generation of wireless networks. Specifically, in Sect. 2.1 we explored UAVs operating as relays and reviewed the relevant research, where we discussed the potential benefits of mmWave communications, while highlighting the significant attenuation in this frequency band. Following this, in Sect. 2.2 we reviewed various beamforming techniques in the realm of mMIMO to overcome the problems associated with mmWave band communications. In this section, we proposed the HBF architecture as a suitable beamforming approach for mmWave UAV-assisted communications and provided the literature review. Then, in Sect. 2.3, we discussed AF and DF relaying schemes in mMIMO technology and reviewed them in the UAV-assisted communications literature. Consequently, in Sect. 2.5 we presented different AI-based algorithms—specifically NI, SL, RL, and DL, with an extensive literature survey on their applications in UAV-assisted communications.

The literature review motivated the research topics presented in the book. In particular, we present the chapters of this book into two main parts:

- Part (I): This part discusses various PSO-based algorithmic solutions for UAV-assisted mMIMO communications. Particularly, Chap. 3 presents two joint beamforming and UAV positioning problems in SU-mMIMO systems with different relaying schemes. Chapter 4 extends the problem to MU-mMIMO systems, while proposing various PSO-based optimization solutions jointly with HBF design followed by Chap. 5 discussing multiple UAV cooperative relaying for enhanced coverage and capacity.
- Part (II): This part mainly focuses on low-complexity DL-based solutions for real-time applications, including Chaps. 6–8. Specifically, Chap. 6 presents a joint UAV location and PA optimization using the DSL to overcome the high computational complexity of the proposed algorithms in Chap. 4. Chapter 7 proposes a DRL-based UAV deployment in a dynamic environment. Finally, Chap. 8 extends the work in Chap. 7 by introducing a novel DRL codebook-based UAV deployment and trajectory design in a dynamic environment to reduce the complexity.

Finally, Chap. 9 concludes this book with conclusions and potential future works.

References

1. K.P. Valavanis, G.J. Vachtsevanos, *Handbook of Unmanned Aerial Vehicles* (Springer, Berlin, 2014), ISBN: 9048197066
2. R. Austin, *Unmanned Aircraft Systems: UAVs Design, Development and Deployment* (Wiley, London, 2011). https://api.semanticscholar.org/CorpusID:109300143
3. R.W. Beard, T.W. McLain, *Small Unmanned Aircraft: Theory and Practice* (Princeton University. Press, Princeton, 2012)

4. M. Asadpour, B. Van den Bergh, D. Giustiniano, K.A. Hummel, S. Pollin, B. Plattner, Micro aerial vehicle networks: an experimental analysis of challenges and opportunities. IEEE Commun. Mag. **52**(7), 141–149 (2014)
5. M. Mozaffari, W. Saad, M. Bennis, M. Debbah, Mobile unmanned aerial vehicles (UAVs) for energy-efficient Internet of Things communications. IEEE Trans. Wireless Commun. **16**(11), 7574–7589 (2017)
6. R.I. Bor-Yaliniz, A. El-Keyi, H. Yanikomeroglu, Efficient 3-D placement of an aerial base station in next generation cellular networks, in *Proceedings of 2016 IEEE International Conference on Communications (ICC)* (2016), pp. 1–5
7. M. Mozaffari, W. Saad, M. Bennis, M. Debbah, Unmanned aerial vehicle with underlaid device-to-device communications: performance and tradeoffs. IEEE Trans. Wireless Commun. **15**(6), 3949–3963 (2016)
8. A. Al-Hourani, S. Kandeepan, S. Lardner, Optimal LAP altitude for maximum coverage. IEEE Wirel. Commun. Lett. **3**(6), 569–572 (2014)
9. M. Mozaffari, W. Saad, M. Bennis, M. Debbah, Efficient deployment of multiple unmanned aerial vehicles for optimal wireless coverage. IEEE Commun. Lett. **20**(8), 1647–1650 (2016)
10. I. Bucaille, S. He´thuin, A. Munari, R. Hermenier, T. Rasheed, S. Allsopp, Rapidly deployable network for tactical applications: aerial base station with opportunistic links for unattended and temporary events absolute example, in *Proceedings of IEEE Military Communications Conference (MILCOM)* (2013), pp. 1116–1120
11. Q. Wu, J. Xu, R. Zhang, UAV-enabled aerial base station (BS) III/III: capacity characterization of UAV-enabled two-user broadcast channel (2018). arvix
12. Q. Wu, R. Zhang, Common throughput maximization in UAV-enabled OFDMA systems with delay consideration. IEEE Trans. Commun. **66**(12), 6614–6627 (2018)
13. M. Mozaffari, W. Saad, M. Bennis, Y.-H. Nam, M. Debbah, A tutorial on UAVs for wireless networks: applications, challenges, and open problems. IEEE Commun. Surv. Tutorials **21**(3), 2334–2360 (2019)
14. K. Meng, D. Li, X. He, M. Liu, Space pruning based time minimization in delay constrained multi-task UAV-based sensing. IEEE Trans. Veh. Technol. **70**(3), 2836–2849 (2021)
15. J. Lyu, Y. Zeng, R. Zhang, Cyclical multiple access in UAV-aided communications: a throughput-delay tradeoff. IEEE Wireless Commun. Lett. **5**(6), 600–603 (2016)
16. Z. Xiao, P. Xia, X.-g. Xia, Enabling UAV cellular with millimeter-wave communication: potentials and approaches. IEEE Commun. Mag. **54**(5), 66–73 (2016)
17. C.A. Balanis, *Antenna Theory: Analysis and Design* (John Wiley & Sons, London, 2016)
18. N. Fatema, G. Hua, Y. Xiang, D. Peng, I. Natgunanathan, Massive MIMO linear precoding: a survey. IEEE Syst. J. **12**(4), 3920–3931 (2017)
19. B. Yang, Z. Yu, J. Lan, R. Zhang, J. Zhou, W. Hong, Digital beamforming-based massive MIMO transceiver for 5G millimeter-wave communications. IEEE Trans. Microw. Theory Techn. **66**(7), 3403–3418 (2018)
20. E. Ali, M. Ismail, R. Nordin, N.F. Abdulah, Beamforming techniques for massive MIMO systems in 5G: overview, classification, and trends for future research. Front. Inf. Technol. Electron. Eng. **18**, 753–772 (2017)
21. M. Mahmood, A. Koc, T. Le-Ngoc, Energy-efficient MU-massive-MIMO hybrid precoder design: low-resolution phase shifters and digital-to-analog converters for 2D antenna array structures. IEEE Open J. Commun. Soc. **2**, 1842–1861 (2021)
22. S.S. Ioushua, Y.C. Eldar, A family of hybrid analog–digital beamforming methods for massive MIMO systems. IEEE Trans. Signal Process. **67**(12), 3243–3257 (2019)
23. Y. Wang, W. Zou, Y. Tao, Analog precoding designs for millimeter wave communication systems. IEEE Trans. Veh. Technol. **67**(12), 11733–11745 (2018)
24. A.F. Molisch et al., Hybrid beamforming for massive MIMO: a survey. IEEE Commun. Mag. **55**(9), 134–141 (2017)
25. X. Gao, L. Dai, S. Han, I. Chih-Lin, R.W. Heath, Energy-efficient hybrid analog and digital precoding for mmWave MIMO systems with large antenna arrays. IEEE J. Sel. Areas Commun. **34**(4), 998–1009 (2016)

26. L. Liang, W. Xu, X. Dong, Low-complexity hybrid precoding in massive multiuser MIMO systems. IEEE Wireless Commun. Lett. **3**(6), 653–656 (2014)
27. T.E. Bogale, L.B. Le, A. Haghighat, L. Vandendorpe, On the number of RF chains and phase shifters, and scheduling design with hybrid analog–digital beamforming. IEEE Trans. Wirel. Commun. **15**(5), 3311–3326 (2016)
28. W. Ni, X. Dong, W. Lu, Near-optimal hybrid processing for massive MIMO systems via matrix decomposition. IEEE Trans. Signal Process. **65**(15), 3922–3933 (2017)
29. S. He, C. Qi, Y. Wu, Y. Huang, Energy-efficient transceiver design for hybrid sub-array architecture MIMO systems. IEEE Access **4**, 9895–9905 (2016)
30. W. Ni, X. Dong, Hybrid block diagonalization for massive multiuser MIMO systems. IEEE Trans. Commun. **64**(1), 201–211 (2016)
31. A. Li, C. Masouros, Hybrid analog-digital millimeter-Wave MU-MIMO transmission with virtual path selection. IEEE Commun. Lett. **21**(2), 438–441 (2017)
32. X. Yu, J.-C. Shen, J. Zhang, K.B. Letaief, Alternating minimization algorithms for hybrid precoding in millimeter wave MIMO systems. IEEE J. Sel. Topics Signal Process. **10**(3), 485–500 (2016)
33. L. Zhu, J. Zhang, Z. Xiao, X.-G. Xia, R. Zhang, Multi-UAV aided millimeter-wave networks: positioning, clustering, and beamforming. IEEE Trans. Wirel. Commun. **21**(7), 4637–4653 (2022)
34. J. Du, W. Xu, Y. Deng, A. Nallanathan, L. Vandendorpe, Energy-saving UAV-assisted multiuser communications with massive MIMO hybrid beamforming. IEEE Commun. Lett. **24**(5), 1100–1104 (2020)
35. X. Pang, J. Tang, N. Zhao, X. Zhang, Y. Qian, Energy-efficient design for mmWave-enabled NOMA-UAV networks. Sci. China Inf. Sci. **64**, 1–14 (2021)
36. W. Feng et al., Hybrid beamforming design and resource allocation for UAV-aided wireless-powered mobile edge computing networks with NOMA. IEEE J. Sel. Areas Commun. **39**(11), 3271–3286 (2021)
37. K. Liu, Y. Liu, P. Yi, Z. Xiao, X.-G. Xia, Deployment and robust hybrid beamforming for UAV MmWave communications. IEEE Trans. Commun. **71**(5), 3073–3086 (2023)
38. X. Yu, X. Huang, K. Wang, F. Shu, X. Dang, Joint design of power allocation, beamforming, and positioning for energy-efficient UAV-aided multiuser millimeter-wave systems. IEEE J. Sel. Areas Commun. **40**(10), 2930–2945 (2022)
39. W. Feng et al., Joint 3D trajectory and power optimization for UAV-aided mmWave MIMO-NOMA networks. IEEE Trans. Commun. **69**(4), 2346–2358 (2021)
40. M. Fozooni, M. Matthaiou, S. Jin, G.C. Alexandropoulos, Massive MIMO relaying with hybrid processing, in Proceedings of IEEE International Conference on Communications (ICC) (IEEE, 2016), pp. 1–6
41. L. Sanguinetti, A.A. D'Amico, Y. Rong, A tutorial on the optimization of amplify-and-forward MIMO relay systems. IEEE J. Sel. Areas Commun. **30**(8), 1331–1346 (2012)
42. K.T. Truong, P. Sartori, R.W. Heath, Cooperative algorithms for MIMO amplify-and-forward relay networks. IEEE Trans. Signal Process. **61**(5), 1272–1287 (2013)
43. K. Xiong, P. Fan, Z. Xu, H.-C. Yang, K.B. Letaief, Optimal cooperative beamforming design for MIMO decode-and-forward relay channels. IEEE Trans. Signal Process. **62**(6), 1476–1489 (2014)
44. S. Zhang, H. Zhang, Q. He, K. Bian, L. Song, Joint trajectory and power optimization for UAV relay networks. IEEE Commun. Lett. **22**(1), 161–164 (2018)
45. Y. Chen, W. Feng, G. Zheng, Optimum placement of UAV as relays. IEEE Commun. Lett. **22**(2), 248–251 (2017)
46. X. Jiang, Z. Wu, Z. Yin, Z. Yang, Power and trajectory optimization for UAV-enabled Amplify-and-Forward relay networks. IEEE Access **6**, 48688–48696 (2018)
47. F. Ono, H. Ochiai, R. Miura, A wireless relay network based on unmanned aircraft system with rate optimization. IEEE Trans. Wirel. Commun. **15**(11), 7699–7708 (2016)
48. Y. Guo, S. Yin, J. Hao, Joint placement and resources optimization for multi-user UAV-relaying systems with underlaid cellular networks. IEEE Trans. Veh. Technol. **69**(10), 12374–12377 (2020)

49. W. Wang, N. Qi, L. Jia, C. Li, T.A. Tsiftsis, M. Wang, Energy-efficient UAV-relaying 5G/6G spectrum sharing networks: interference coordination with power management and trajectory design. IEEE Open J. Commun. Soc. **3**, 1672–1687 (2022)
50. L. Sboui, H. Ghazzai, Z. Rezki, M.-S. Alouini, Achievable rates of UAV-relayed cooperative cognitive radio MIMO systems. IEEE Access **5**, 5190–5204 (2017)
51. W. Tong, G.Y. Li, Nine challenges in artificial intelligence and wireless communications for 6g. IEEE Wirel. Commun. **29**(4), 140–145 (2022)
52. S. Rezwan, W. Choi, Artificial intelligence approaches for UAV navigation: recent advances and future challenges. IEEE Access **10**, 26320–26339 (2022)
53. M.-A. Lahmeri, M.A. Kishk, M.-S. Alouini, Artificial intelligence for UAV-enabled wireless networks: a survey. IEEE Open J. Commun. Soc. **2**, 1015–1040 (2021)
54. S. Sai, A. Garg, K. Jhawar, V. Chamola, B. Sikdar, A comprehensive survey on artificial intelligence for unmanned aerial vehicles. IEEE Open J. Veh. Technol. **4**, 713–738 (2023)
55. W. Tong, A. Hussain, W.X. Bo, S. Maharjan, Artificial intelligence for vehicle-to-everything: a survey. IEEE Access **7**, 10823–10843 (2019)
56. J. Chen, U. Yatnalli, D. Gesbert, Learning radio maps for UAV-aided wireless networks: a segmented regression approach, in *Proceedings of 2017 IEEE International Conference on Communications (ICC)* (2017), pp. 1–6
57. Q. Zhang, W. Saad, M. Bennis, X. Lu, M. Debbah, W. Zuo, Predictive deployment of UAV base stations in wireless networks: machine learning meets contract theory. IEEE Trans. Wirel. Commun. **20**(1), 637–652 (2021)
58. K. Xiao, J. Zhao, Y. He, S. Yu, Trajectory prediction of UAV in Smart City using recurrent neural networks, in *Proceedings of IEEE International Conference on Communications (ICC)* (2019), pp. 1–6
59. J. Kwak, Y. Sung, Autoencoder-based candidate waypoint generation method for autonomous flight of multi-unmanned aerial vehicles. Adv. Mech. Eng. **11**(6), 1687814019856772 (2019)
60. R.S. Sutton, A.G. Barto, *Reinforcement Learning: An Introduction* (MIT Press, Cambridge, 2018). ISBN: 0262039249
61. Y. Liao, G. Yu, P. Chen, B. Zhou, H. Li, Integration of decision-making and motion planning for autonomous driving based on double-layer reinforcement learning framework. IEEE Trans. Veh. Technol. **73**(3), 3142–3158 (2024)
62. M. Al-Sharman, R. Dempster, M.A. Daoud, M. Nasr, D. Rayside, W. Melek, Self-learned autonomous driving at unsignalized intersections: a hierarchical reinforced learning approach for feasible decision-making. IEEE Trans. Intell. Transp. Syst. **24**(11), 12345–12356 (2023)
63. Y. Hu, W. Wang, H. Liu, L. Liu, Reinforcement learning tracking control for robotic manipulator with kernel-based dynamic model. IEEE Trans. Neural Netw. Learn. Syst. **PP**, 1–9 (2019)
64. Z. Hou, J. Fei, Y. Deng, J. Xu, Data-efficient hierarchical reinforcement learning for robotic assembly control applications. IEEE Trans. Ind. Electron. **68**(11), 11565–11575 (2021)
65. A. Hajijamali Arani, M.M. Azari, P. Hu, Y. Zhu, H. Yanikomeroglu, S. Safavi-Naeini, Reinforcement learning for energy-efficient trajectory design of UAVs. IEEE Internet Things J. **9**(11), 9060–9070 (2022)
66. H. Dong, D. Zhao, D. Huang, K. Yan, W. Ren, SARSA (λ) reinforcement learning based path planning of unmanned aerial vehicles, in *Proceedings of 2024 IEEE 13th Data-Driven Control Learning System Conference (DDCLS)* (2024), pp. 1099–1105
67. M. Muzammul, M. Assam, Y.Y. Ghadi, N. Innab, M. Alajmi, T.J. Alahmadi, IR-QLA: machine learning-based Q-learning algorithm optimization for UAVs faster trajectory planning by instructed-reinforcement learning. IEEE Access **12**, 91300–91315 (2024)
68. I. Zaghbani, R. Jarray, S. Bouallègue, Comparative study of Q-learning and SARSA algorithms for UAV path planning in 3D environments, in *Proceedings of 2024 IEEE 28th International Journal of Intelligent Engineering and Systems (INES)* (2024), pp. 000245–000250
69. N.Q.H. Othman, J. Li, Q. Yang, Multi-agent reinforcement learning aided resource allocation with SARSA in UAV networks, in *Proceedings of 2023 IEEE International Conference on Signal Processing, Communications and Computing (ICSPCC)* (2023), pp. 1–6

70. Y. LeCun, Y. Bengio, G. Hinton, Deep learning. Nature **521**, 436–44 (2015)
71. H. Huang et al., Deep learning for physical-layer 5G wireless techniques: opportunities, challenges and solutions. IEEE Wirel. Commun. **27**(1), 214–222 (2020)
72. Y. Sun, M. Peng, Y. Zhou, Y. Huang, S. Mao, Application of machine learning in wireless networks: key techniques and open issues. IEEE Commun. Surv. Tuts. **21**(4), 3072–3108 (2019)
73. X. Zhu, A. Koc, R. Morawski, T. Le-Ngoc, A deep learning and geospatial data-based channel estimation technique for hybrid massive MIMO systems. IEEE Access **9**, 145115–145132 (2021)
74. M.A. Nielsen, *Neural Networks and Deep Learning*, vol. 25 (Determination Press, San Francisco, 2015)
75. Y. LeCun, Y. Bengio, G. Hinton, Deep learning. Nature **521**(7553), 436–444 (2015)
76. V. Sze, Y.-H. Chen, T.-J. Yang, J.S. Emer, Efficient processing of deep neural networks: a tutorial and survey. Proc. IEEE **105**(12), 2295–2329 (2017)
77. W. Samek, G. Montavon, S. Lapuschkin, C.J. Anders, K.-R. Müller, Explaining deep neural networks and beyond: a review of methods and applications. Proc. IEEE **109**(3), 247–278 (2021)
78. K. O'Shea, R. Nash, An introduction to convolutional neural networks (2015). ArXiv e-prints
79. J. Wu, *Introduction to Convolutional Neural Networks*, vol. 5(23) (National Key Lab Novel Software Technology., Nanjing University, China, 2017), p. 495
80. Z. Li, F. Liu, W. Yang, S. Peng, J. Zhou, A survey of convolutional neural networks: analysis, applications, and prospects. IEEE Trans. Neural Networks Learn. Syst. **33**(12), 6999–7019 (2022)
81. I. Sutskever, O. Vinyals, Q.V. Le, Sequence to sequence learning with neural networks, in *Advances in Neural Information Processing Systems*, vol. 27 (2014)
82. Z.C. Lipton, J. Berkowitz, C. Elkan, A critical review of recurrent neural networks for sequence learning (2015). arXiv preprint, vol. arXiv:1506.00019
83. X. Zhu, F. Qi, Y. Feng, Deep-learning-based multiple beamforming for 5G UAV IoT networks. IEEE Netw. **34**(5), 32–38 (2020)
84. F. Jiang, K. Wang, L. Dong, C. Pan, W. Xu, K. Yang, Deep-learning-based joint resource scheduling algorithms for hybrid MEC networks. IEEE Internet Things J. **7**(7), 6252–6265 (2020)
85. Y. Pan, Y. Yang, W. Li, A deep learning trained by genetic algorithm to improve the efficiency of path planning for data collection with multi-UAV. IEEE Access **9**, 7994–8005 (2021)
86. H. Huang, Y. Yang, H. Wang, Z. Ding, H. Sari, F. Adachi, Deep reinforcement learning for UAV navigation through massive MIMO technique. IEEE Trans. Veh. Technol. **69**(1), 1117–1121 (2020)
87. S.P. Gopi, M. Magarini, Reinforcement learning aided UAV base station location optimization for rate maximization. Electronics **10**(23), (2021). ISSN: 2079-9292. https://www.mdpi.com/2079-9292/10/23/2953
88. C. Wang, J. Wang, J. Wang, X. Zhang, Deep-reinforcement-learning-based autonomous UAV navigation with sparse rewards. IEEE Internet Things J. **7**(7), 6180–6190 (2020)
89. C. Zhang, Z. Li, C. He, K. Wang, C. Pan, Deep reinforcement learning based trajectory design and resource allocation for UAV-assisted communications. IEEE Commun. Lett. **27**(9), 2398–2402 (2023)
90. H. Wang, H. Zhang, X. Liu, K. Long, A. Nallanathan, Joint UAV placement optimization, resource allocation, and computation offloading for THz band: a DRL approach. IEEE Trans. Wirel. Commun. **22**(7), 4890–4900 (2023)
91. S. Khairy, P. Balaprakash, L. X. Cai, Y. Cheng, Constrained deep reinforcement learning for energy sustainable multi-UAV based random access IoT networks with NOMA. IEEE J. Sel. Areas Commun. **39**(4), 1101–1115 (2021)

Part I
PSO-Based Algorithmic Solutions for UAV-Assisted mMIMO Communications

Chapter 3
Joint HBF and UAV Deployment in Dual-Hop SU-mMIMO Systems

3.1 Introduction

In the previous chapter, we provided an extensive review of the recent research on UAV-assisted communications. Most prior research on the use of UAVs to assist in communications has overlooked the potential benefits of beamforming solutions, as evidenced by studies such as [1–7]. While some studies have explored beamforming solutions, they have been limited to single-hop communications scenarios and have only examined its application in single links, such as in studies [8–14], where UAVs act as flying BSs rather than relays, making the beamforming solutions inapplicable for dual-hop or relaying structures. The optimization of UAV location and beamforming design for a dual-hop mmWave SU-mMIMO to maximize the total AR is an unaddressed problem, presenting a significant opportunity to advance the field of UAV-based wireless communications. To address this, we present various joint beamforming and UAV positioning optimizations. Specifically, we consider two main mmWave dual-hop SU-mMIMO UAV-relaying scenarios as follows:

- We assume a system model in which a UAV acts as an AF relay between a BS and the gateway. Then, we propose a novel joint optimization scheme to maximize the end-to-end throughput in this UAV-assisted mmWave SU-mMIMO communications systems using two different spherical array (SA) configurations, namely SA equal angle (SA-EA) and SA uniform space (SA-US) as shown in Fig. 3.1. The challenging non-convex optimization problem is solved using PSO method to find optimal UAV relay positioning within a given deployment region. Then, HBFs for the BS and gateway are designed using an orthogonal matching pursuit (OMP)-based algorithmic solution, whereas, for the UAV relay, considering ABF, both transmit and receive RF beamformers are formulated using SVD of the channel matrices. The illustrative results show the

T. Le-Ngoc et al., *AI-Enabled UAV-Assisted Massive MIMO*, Wireless Networks,
https://doi.org/10.1007/978-3-032-07872-8_3

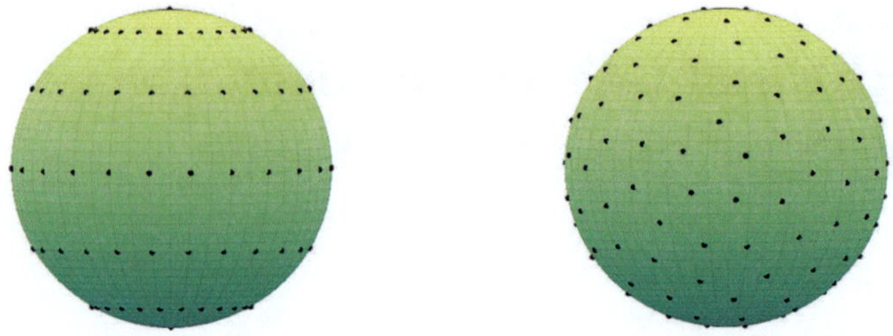

(a) (b)

Fig. 3.1 Spherical array configurations: (**a**) equal angle (EA) and (**b**) uniform space (US)

improved spectral and energy efficiencies achieved for two SA configurations when compared to uniform rectangular array (URA).

- We consider a similar system model with BS, UAV, and the gateway, while this time the UAV acts as a DF relay and uses HBF setup. Specifically, we extend the problem by considering HBF design for UAV relay, in addition to the BS and gateway, jointly with UAV location optimization. Similarly, we use PSO to optimize the UAV location. Then, based the optimized UAV location, we design HBF for the BS, UAV, and the gateway, with (i) the RF stages utilizing slow time-varying angle of departure (AoD)/angle of arrival (AoA) information and (ii) baseband (BB) stages using the reduced-dimensional effective channel matrices. Illustrative results show that the proposed HBF can achieve higher energy efficiency compared to FDBF, while finding near-optimal UAV locations within the given deployment range.

The remainder of this chapter is organized as follows. Section 3.2 introduces the channel model for UAV-assisted mmWave SU-mMIMO systems. Section 3.3 describes the system model in which the UAV operates as a DF relay using an ABF configuration, and presents the corresponding illustrative results. Building on this, Sect. 3.4 examines a similar setup, this time employing a HBF configuration at the UAV, and provides the associated numerical results. Finally, Sect. 3.5 concludes the chapter.

3.2 Channel Model

We consider mmWave channels for both BS-UAV and UAV-gateway links. While LoS channel models can be useful for simple scenarios, they can be limited in their ability to capture the channel complexities (e.g., multipath fading and shadowing). Thus, mmWave channel models can provide a more accurate representation of the channel characteristics, including the impact of none-LoS (NLoS) paths and obstacles on the signal propagation in UAV-assisted mMIMO communications. Let (x_u, y_u, z_u), (x_b, y_b, z_b), and $\left(x_g, y_g, z_g\right)$ denote the locations of UAV, BS, and gateway, respectively. Based on the Saleh-Valenzuela channel model [15], the channel is given as follows:

$$\mathbf{H}_i = \sum_{l=1}^{L} z_{l,i} \tau_i^{-\eta} \boldsymbol{\phi}_r \left(\theta_{r,l,i}, \psi_{r,l,i}\right) \boldsymbol{\phi}_t^T \left(\theta_{r,l,i}, \psi_{r,l,i}\right), \tag{3.1}$$

where L is the total number of paths, η is the path loss exponent, and $z_{l,i}$ is the complex gain of lth path in ith channel, where $i = \{1, 2\}$. $\boldsymbol{\phi}(.,.)$ is the Tx/Rx array steering vector for URA, which is given as [16]:

$$\begin{aligned}\boldsymbol{\phi}_{(j)}(\theta, \psi) =& [1, e^{-j2\pi d \sin(\theta)\cos(\psi)}, \cdots, e^{-j2\pi d (N_x - 1)\sin(\theta)\cos(\psi)}] \\ & \otimes [1, e^{-j2\pi d \sin(\theta)\sin(\psi)}, \cdots, e^{-j2\pi d (N_y - 1)\sin(\theta)\sin(\psi)}],\end{aligned} \tag{3.2}$$

where $j = \{t, r\}$, d is the inter-element spacing, and N_x $\left(N_y\right)$ is the horizontal (vertical) size of the array. Here, the angles $\theta_{r,l,i} \in \left[\theta_i - \delta_i^\theta, \theta_i + \delta_i^\theta\right]$ and $\psi_{r,l,i} \in \left[\psi_i - \delta_i^\psi, \psi_i + \delta_i^\psi\right]$ are the elevation angle of arrival (EAoA) and azimuth angle of arrival (AAoA) for the lth path in the ith channel, respectively. Similarly, the angles $\theta_{t,l,i} \in \left[\theta_i - \delta_i^\theta, \theta_i + \delta_i^\theta\right]$ and $\psi_{t,l,i} \in \left[\psi_i - \delta_i^\psi, \psi_i + \delta_i^\psi\right]$ are the elevation angle of departure (EAoD) and azimuth angle of departure (AAoD). θ_i (ψ_i) is the mean EAoD (AAoD)/EAoA (AAoA) with angular spread δ_i^θ $\left(\delta_i^\psi\right)$. Consequently, τ_1 and τ_2 are defined as

$$\tau_1 = \sqrt{(x_b - x_u)^2 + (y_b - y_u)^2 + (z_b - z_u)^2}, \tag{3.3}$$

$$\tau_2 = \sqrt{\left(x_u - x_g\right)^2 + \left(y_u - y_g\right)^2 + \left(z_u - z_g\right)^2}. \tag{3.4}$$

This channel model will be used for the system models in Sects. 3.3 and 3.4.

3.3 Joint HBF and UAV Deployment in Dual-Hop SU-mMIMO Systems: AF Relaying

In this section, we provide the first system model, where the UAV acts as an AF relay, while considering ABF at the UAV, with the purpose of maximizing the total AR from the BS to the gateway.

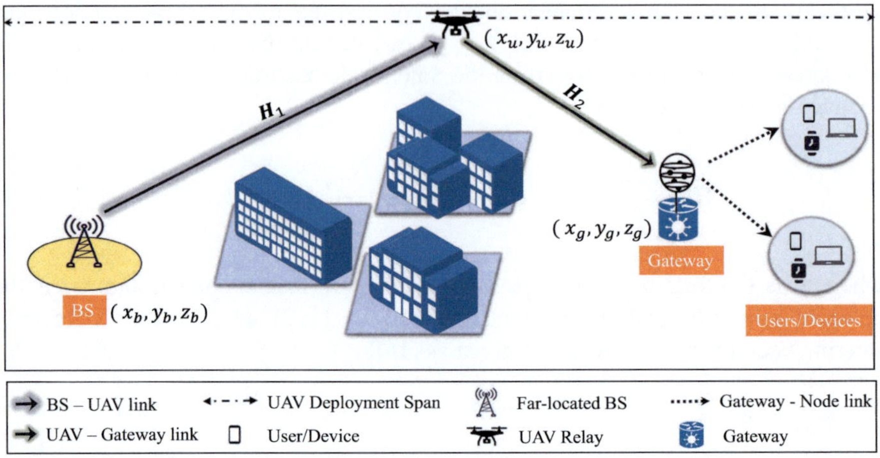

Fig. 3.2 UAV-assisted SU-mMIMO communications system

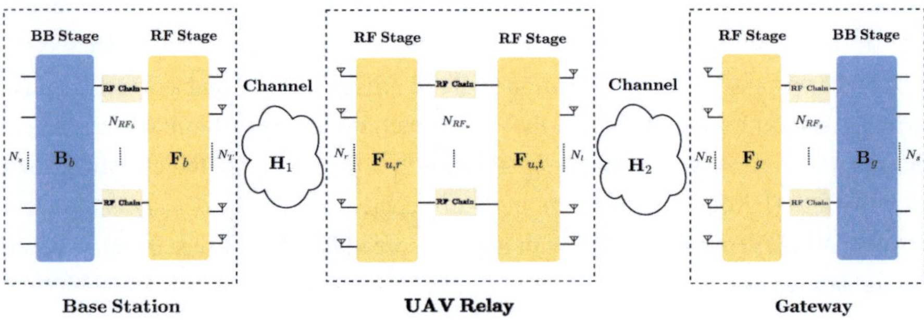

Fig. 3.3 UAV-assisted SU-mMIMO AF relaying architecture

3.3.1 System Model

We consider users/devices and the gateway are located in a remote area, which is difficult to access directly by BS (eNodeB) due to obstacles, and a UAV is used as an AF relay to improve the situation as shown in Fig. 3.2. We assume the UAV relay is placed at a height $z_u > z_b$, where z_b is the height of the BS.

In the system model as shown in Fig. 3.3, we consider the BS is equipped with N_T transmit antennas, UAV relay with $N_r(N_t)$ antennas for receiving (transmitting) signals, and gateway with N_R antennas. Both BS and gateway adopt HBF architecture. Here, BS consists of a RF beamforming stage $\mathbf{F}_b \in \mathbb{C}^{N_T \times N_{RF_b}}$ and BB stage $\mathbf{B}_b$ $\mathbb{C}^{N_{RF_b} \times N_s}$, where N_s represents the data streams from the BS and N_{RF_b} is the RF chains such that $N_s \leq N_{RF_b} \leq N_T$ to guarantee multistream transmission. Considering half-duplex (HD) UAV AF relaying, BS sends N_s data streams through channel $\mathbf{H}_1 \in \mathbb{C}^{N_r \times N_T}$ in the first time slot. Using N_r antennas, the UAV relay receives signals with an RF combiner $\mathbf{F}_{u,r} \in \mathbb{C}^{N_{RF_u} \times N_r}$. We assume UAV relay transmits the data in the second time slot using RF beamformer $\mathbf{F}_{u,t} \in \mathbb{C}^{N_t \times N_{RF_u}}$ through channel $\mathbf{H}_2 \in \mathbb{C}^{N_R \times N_t}$. The received signal at gateway is processed through RF stage $\mathbf{F}_g \in \mathbb{C}^{N_{RF_g} \times N_R}$ and BB stage $\mathbf{B}_g$ $\mathbb{C}^{N_s \times N_{RF_g}}$. Here, all RF beamforming and combining stages are implemented using PSs and thus impose a CM constraint, i.e., $|\mathbf{F}_b(i,j)| = \frac{1}{\sqrt{N_T}}$, $(|\mathbf{F}_g(i,j)| = \frac{1}{\sqrt{N_R}})$, $|\mathbf{F}_{u,r}(i,j)| = \frac{1}{\sqrt{N_r}}$, and $|\mathbf{F}_{u,t}(i,j)| = \frac{1}{\sqrt{N_t}}$ $\forall i, j$. The use of RF beamforming and combining stages for BS, UAV relay, and gateway greatly reduces the number of RF chains from: (1) N_T to N_{RF_u}; (2) $N_r(N_t)$ to N_{RF_u}; and (3) N_R to N_{RF_g}, respectively. Considering the data signal is $\mathbf{d} = [d_1, d_2, \ldots, d_{N_s}]^T$ with $\mathbb{E}\{\mathbf{d}\mathbf{d}^H\} = \mathbf{I}_{N_s} \in \mathbb{C}^{N_s \times N_s}$, then the signal transmitted by BS can be written as follows:

$$\mathbf{s} = \mathbf{F}_b \mathbf{B}_b \mathbf{d}. \tag{3.5}$$

The power constraint at BS can be expressed as $\|\mathbf{F}_1\mathbf{B}_1\|_F^2 = P_1$, where P_1 denotes the total transmit power of BS. Then, the received signal at the UAV relay is given as follows:

$$\begin{aligned} \mathbf{y}_{u,r} &= \mathbf{H}_1 \mathbf{s} + \mathbf{n}_1, \\ &= \mathbf{H}_1 \mathbf{F}_b \mathbf{B}_b \mathbf{d} + \mathbf{n}_1, \end{aligned} \tag{3.6}$$

where $\mathbf{n}_1 \in \mathbb{C}^{N_r}$ denotes the zero-mean complex circularly symmetric Gaussian noise vector at UAV relay with covariance matrix $\mathbb{E}\{\mathbf{n}_1\mathbf{n}_1^H\} = \sigma_1^2 \mathbf{I}_{N_r} \in \mathbb{C}^{N_r \times N_r}$. Then, the received signal after RF combining at the UAV relay is written as follows:

$$\tilde{\mathbf{y}}_{u,r} = \mathbf{F}_{u,r}\mathbf{H}_1\mathbf{F}_b\mathbf{B}_b\mathbf{d} + \mathbf{F}_{u,r}\mathbf{n}_1. \tag{3.7}$$

Ignoring the delay constraints, the transmit signal at the UAV relay is given as follows:

$$\mathbf{y}_{u,t} = \mathbf{F}_{u,t}\mathbf{F}_{u,r}\mathbf{H}_1\mathbf{F}_b\mathbf{B}_b\mathbf{d} + \mathbf{F}_{u,t}\mathbf{F}_{u,r}\mathbf{n}_1. \tag{3.8}$$

The power constraint of the transmitted signal at the UAV relay can be expressed as $\|\mathbf{F}_{u,t}\mathbf{F}_{u,r}\|_F^2 = P_u$, where P_u denotes the total transmit power of the UAV. The received signal at gateway can be expressed as follows:

$$\mathbf{y}_d = \mathbf{H}_2\mathbf{F}_{u,t}\mathbf{F}_{u,r}\mathbf{H}_1\mathbf{F}_b\mathbf{B}_b\mathbf{d} + \mathbf{H}_2\mathbf{F}_{u,t}\mathbf{F}_{u,r}\mathbf{n}_1 + \mathbf{n}_2, \tag{3.9}$$

where $\mathbf{n}_2$ denotes the additive circular symmetric Gaussian noise such that $\mathbf{n}_2 \sim \mathcal{CN}(\mathbf{0}, \sigma_2^2\mathbf{I}_{N_R})$. After HBF at the gateway, the received signal can be written as follows:

$$\tilde{\mathbf{y}}_d = \underbrace{\mathbf{U}_g\mathbf{H}_2\mathbf{F}_{u,t}\mathbf{F}_{u,r}\mathbf{H}_1\mathbf{W}_b\mathbf{d}}_{\text{Desired Signal}} + \underbrace{\mathbf{U}_g\mathbf{H}_2\mathbf{F}_{u,t}\mathbf{F}_{u,r}\mathbf{n}_1}_{\text{UAV Relay Noise}} + \underbrace{\mathbf{U}_g\mathbf{n}_2}_{\text{Gateway Noise}}, \tag{3.10}$$

where $\mathbf{U}_b = \mathbf{F}_b\mathbf{B}_b$ and $\mathbf{U}_g = \mathbf{B}_g\mathbf{F}_g$. Then, the spectral efficiency can be calculated as

$$
\begin{aligned}
\mathrm{R} = &\frac{1}{2}\log_2|\mathbf{I}_{N_s} + \cdots \\
&\frac{1}{N_s}\mathbf{Q}_n^{-1}\left(\sqrt{P_u P_b}\mathbf{U}_g\mathbf{H}_2\mathbf{F}_{u,t}\mathbf{F}_{u,r}\mathbf{H}_1\mathbf{U}_b\right)\left(\sqrt{P_u P_b}\mathbf{U}_g\mathbf{H}_2\mathbf{F}_{u,t}\mathbf{F}_{u,r}\mathbf{H}_1\mathbf{U}_b\right)^H|,
\end{aligned}
\tag{3.11}
$$

where $\mathbf{Q}_n^{-1} = \sigma_n^2\left[(\sqrt{P_u}\mathbf{U}_g\mathbf{H}_2\mathbf{F}_{u,t}\mathbf{F}_{u,r})(\sqrt{P_u}\mathbf{U}_g\mathbf{H}_2\mathbf{F}_{u,t}\mathbf{F}_{u,r})^H + \mathbf{U}_g\mathbf{U}_g^H\right]^{-1}$ is the covariance matrix of Gaussian noise at UAV relay.

3.3.2 Problem Formulation

For UAV-assisted SU-mMIMO communications system, where the UAV is deployed at a fixed height z_u and relaying data to the gateway, the total AR can be maximized by the joint optimization of $\mathbf{F}_b$, $\mathbf{B}_b$, $\mathbf{F}_{u,t}$, $\mathbf{F}_{u,r}$, $\mathbf{F}_g$, $\mathbf{B}_g$, and $\mathbf{x}_u = [x_u, y_u] \in \mathbb{R}^2$, where $\mathbf{x}_u$ represents UAV relay position, which is to be optimized within the given UAV flying span. Then, we can formulate the joint optimization problem as

$$
\begin{aligned}
&\max_{\{\mathbf{F}_b,\mathbf{B}_b,\mathbf{F}_{u,t},\mathbf{F}_{u,r},\mathbf{F}_g,\mathbf{B}_g,x_u,y_u\}} \quad \mathrm{R}(\mathbf{F}_b, \mathbf{B}_b, \mathbf{F}_{u,t}, \mathbf{F}_{u,r}, \mathbf{F}_g, \mathbf{B}_g, x_u, y_u) \\
&\text{s.t. } C_1: |\mathbf{F}_{u,t}(i,j)| = \frac{1}{\sqrt{N_t}}, |\mathbf{F}_{u,r}(i,j)| = \frac{1}{\sqrt{N_r}}, \ \forall i, j, \\
&\quad C_2: |\mathbf{F}_b(i,j)| = \frac{1}{\sqrt{N_T}}, |\mathbf{F}_g(i,j)| = \frac{1}{\sqrt{N_g}}, \ \forall i, j, \\
&\quad C_3: \mathbb{E}\{\|\mathbf{s}\|_2^2\} \le P_1, \mathbb{E}\{\left\|\mathbf{y}_{u,t}\right\|_2^2\} \le P_u, \\
&\quad C_4: x_{\min} \le x_u \le x_{\max}, y_{\min} \le y_u \le y_{\max},
\end{aligned}
\tag{3.12}
$$

where C_1 and C_2 refer to the CM constraints due to the use of PSs for UAV and BS/gateway, respectively. C_3 indicates the transmit power constraint for the UAV relay and BS, and C_4 implies UAV positioning within the given deployment region. Here, $[(x_{\min}, y_{\min}), (x_{\max}; y_{\max})]$ represents the UAV flying span in x-y plane. The optimization problem defined in (3.12) is non-convex and intractable. To solve this problem, we sequentially develop beamforming stages for BS, UAV and gateway based on an arbitrary fixed UAV location to optimize $[x_u, y_u]$ using PSO. Then, based on optimal UAV relay position, we reformulate RF and BB stages for BS and gateway as well as RF beamformer/combiner for UAV based on instantaneous channel matrices $\mathbf{H}_1$ and $\mathbf{H}_2$.

3.3.3 Joint Beamforming and UAV Positioning

To maximize the spectral efficiency (as given in (3.12)), we consider ABF for UAV relay, where we design both transmit and receive RF beamformer/combiner and optimize UAV location together with HBF (i.e., design of RF beamformer and BB stage) for BS and gateway. Since the optimization problem is intractable, we first design $\mathbf{F}_b, \mathbf{B}_b, \mathbf{F}_{u,t}, \mathbf{F}_{u,r}, \mathbf{F}_g, \mathbf{B}_g$ based on an arbitrary fixed UAV location and then reformulate RF beamformers and BB stages for optimal UAV location using PSO. Since SVD-based beamforming design is considered optimal for point-to-point mMIMO systems [17], the design of HBF stages for BS and gateway can be found by minimizing the Frobenius norm of the difference between optimal unconstrained beamformers and the corresponding HBF stages as follows:

$$\begin{aligned} \min_{\{\mathbf{F}_i,\mathbf{B}_i\}} \quad & \|\mathbf{F}_{\text{opt,i}} - \mathbf{F}_i\mathbf{B}_i\|_F \\ \text{s.t.} \quad & C_2, C_3, \end{aligned} \tag{3.13}$$

where $\mathbf{F}_{\text{opt,i}}$ is the unconstrained beamformer (combiner) for $i = \{b, g\}$, which is derived from the SVD of the channel matrix between BS and UAV (between UAV and gateway). Using SVD, $\mathbf{H}_1$ can be written as follows:

$$\mathbf{H}_1 = \mathcal{Q}_1 \boldsymbol{\Sigma}_1 \mathcal{V}_1^H \in \mathbb{C}^{N_r \times N_T}, \tag{3.14}$$

where $\mathcal{Q}_1$ and $\mathcal{V}_1$ are $N_r \times N_r$ and $N_T \times N_T$ unitary matrices, respectively. $\boldsymbol{\Sigma}_1$ is a $N_r \times N_T$ rectangular diagonal matrix with nonnegative real numbers on the diagonal. Then, the unconstrained optimal beamformer for BS is formulated as $\mathbf{F}_{\text{opt,b}} = \mathcal{V}_{1,1}$, where $\mathcal{V}_1 = [\mathcal{V}_{1,1}\mathcal{V}_{1,2}] \in \mathbb{C}^{N_T \times N_T}$ [17]. Using SVD, the channel matrix $\mathbf{H}_2$ can be written as follows:

$$\mathbf{H}_2 = \mathcal{Q}_2 \boldsymbol{\Sigma}_2 \mathcal{V}_2^H \in \mathbb{C}^{N_R \times N_t}. \tag{3.15}$$

Then, the optimal unconstrained combiner for the gateway can be formulated as $\mathbf{F}_{\text{opt,g}} = \mathcal{V}_{2,1}$, i.e., right singular vectors of $\mathbf{H}_2$. The HBF stages $\mathbf{F}_b, \mathbf{B}_b, \mathbf{F}_g, \mathbf{B}_g$ are designed using an OMP-based algorithmic solution, which is outlined in Algorithm 3.1. For UAV relay, we design transmit and receive RF beamformer and combiner based on angular location of UAV and gateway, respectively. Let $\boldsymbol{e}_n = [e^{j\upsilon_{1,n}}, e^{j\upsilon_{2,n}}, \ldots, e^{j\upsilon_{N_{u,r},n}}]^T \in \mathbb{C}^{N_{u,r}}$ be the receive phase vector, where $\upsilon_{m,n}$ is the phase angle of the $(m, n)th$ entry of $\mathcal{Q}_1$. Then, we can formulate the receive phase matrix $\boldsymbol{\Psi}_u^r$ as follows:

$$\boldsymbol{\Psi}_u^r = [\boldsymbol{\psi}_1, \boldsymbol{\psi}_2, \ldots, \boldsymbol{\psi}_{N_{u,r}}] \in \mathbb{C}^{N_{u,r} \times N_{u,r}}. \tag{3.16}$$

Comparing (3.14) and (3.16), we can see that each entry of ψ_n and $Q_1(n)$ has the same phase angle. Using N_{RF_u} RF chains at UAV relay, the receive RF combiner can be formulated as

$$\mathbf{F}_{u,r} = [\boldsymbol{\psi}_1, \boldsymbol{\psi}_2, \ldots, \boldsymbol{\psi}_{N_{RF_u}}]^T \in \mathbb{C}^{N_{RF_u} \times N_{u,r}}. \tag{3.17}$$

Following a similar approach, the transmit RF beamformer at UAV can be formulated by using SVD of $\mathbf{H}_2$. Then, the transmit phase matrix $\boldsymbol{\Psi}_u^t$ can be written as follows:

$$\boldsymbol{\Psi}_u^t = [\boldsymbol{\varphi}_1, \boldsymbol{\varphi}_2, \ldots, \boldsymbol{\varphi}_{N_{u,t}}] \in \mathbb{C}^{N_{u,t} \times N_{u,t}}, \tag{3.18}$$

where $\boldsymbol{\varphi}_n = [e^{j\varphi_{1,n}}, e^{j\varphi_{2,n}}, \ldots, e^{j\varphi_{N_{u,t},n}}]^T \in \mathbb{C}^{N_{u,t}}$ is the transmit phase vector, where $\varphi_{m,n}$ is the phase angle of the $(m,n)th$ entry of $\mathcal{V}_2$. Then, using N_{RF_u} RF chains at the UAV relay, the transmit RF beamformer can be formulated as

$$\mathbf{F}_{u,t} = [\boldsymbol{\varphi}_1, \boldsymbol{\varphi}_2, \ldots, \boldsymbol{\varphi}_{N_{RF_u}}] \in \mathbb{C}^{N_{u,t} \times N_{RF_u}}. \tag{3.19}$$

After the design of RF and BB stages, the optimization problem given in (3.12) can be reformulated as follows:

$$\begin{aligned} &\max_{\{x_o, y_o\}} \quad \mathrm{R}(x_o, y_o) \\ &\text{s.t.} \quad C_4 : x_{\min} \le x_o \le x_{\max},\, y_{\min} \le y_o \le y_{\max}. \end{aligned} \tag{3.20}$$

The resulting optimization problem in (3.20) is still an NP hard and remains a computationally challenging task. Therefore, we propose a PSO-based algorithmic solution to optimize UAV position and sequentially update the RF and BB stages for BS, UAV, and gateway. Initially, Z_p particles are randomly placed in search space, each with its own position, velocity, and fitness value. Then, during tth iteration, the position $\mathbf{X}_i^{(t)}$ and velocity $\mathbf{V}_i^{(t)}$ of each ith particle are updated as follows [18]:

$$\mathbf{X}_i^{(t+1)} = \mathbf{X}_i^{(t)} + \mathbf{V}_i^{(t+1)}, \tag{3.21}$$

$$\mathbf{V}_i^{(t+1)} = \gamma_1 \mathbf{Y}_1^{(t)}(\mathbf{X}_{best}^{(t)} - \mathbf{X}_i^{(t)}) + \gamma_2 \mathbf{Y}_2^{(t)}(\mathbf{X}_{best,i}^{(t)} - \mathbf{X}_i^{(t)}) + \gamma_3 \mathbf{V}_i^{(t)}, \tag{3.22}$$

where γ_1, γ_2, and γ_3 are tuning parameters and $\mathbf{Y}_1^{(t)}$ and $\mathbf{Y}_2^{(t)}$ are the matrices with entries uniformly distributed in [0, 1]. Here, each ith particle represents the candidate position of UAV, and it iteratively updates its personal best $\mathbf{X}_{best,i}^{(t)}$ and global best $\mathbf{X}_{best}^{(t)}$ solution during a total of T iterations as follows:

$$\mathbf{X}_{best,i}^{(t)} = \underset{\mathbf{X}_i^{(t^*)}, \forall t^*=0,1,\cdots,t}{\arg\max} \; \mathrm{R}(\mathbf{F}_1, \mathbf{B}_1, \mathbf{F}_{u,t}, \mathbf{F}_{u,r}, \mathbf{F}_2, \mathbf{B}_2, \mathbf{X}_i^{(t^*)}), \tag{3.23}$$

Algorithm 3.1: Joint beamforming and UAV positioning

1 **Input**: Z_p, T, (θ, ψ), (x_b, y_b, z_b), (x_g, y_g, z_g), (x_u, y_u, z_u).
2 **Output**: $\mathbf{x}$, $\mathbf{F}_b$, $\mathbf{B}_b$, $\mathbf{F}_{u,t}$, $\mathbf{F}_{u,r}$, $\mathbf{F}_g$, $\mathbf{B}_g$.
3 Compute $\mathbf{H}_1$ and $\mathbf{H}_2$ using (3.1).
4 Find the unitary matrices $\mathcal{Q}_1$ ($\mathcal{Q}_2$) and $\mathcal{V}_1$ ($\mathcal{V}_2$) using SVD.
5 **for** $i = \{b, g\}$ **do**
6     $\mathbf{F}_i = [\ ]$, $\mathbf{F}_{res,i} = \mathbf{F}_{opt,i}$.
7   **for** $j = 1 : N_{RF_i}$ **do**
8         $\Upsilon_i = (\mathbf{A}_i)^T \mathbf{F}_{res,i}$.
9   Find the index k which maximizes $(\Upsilon_i \Upsilon_i^*)$.
10         $\mathbf{F}_i = \left[\mathbf{F}_i \mid (\mathbf{A}_i^T)^{(k)}\right]$.
11         $\mathbf{B}_i = \left(\mathbf{F}_i^H \mathbf{F}_i\right)^{-1} \mathbf{F}_i^H \mathbf{F}_{opt,i}$.
12         $\mathbf{F}_{res,i} = \frac{\mathbf{F}_{opt,i} - \mathbf{F}_i \mathbf{B}_i}{\|\mathbf{F}_{opt,i} - \mathbf{F}_i \mathbf{B}_i\|_F}$.
13   **end**
14     $\mathbf{B}_i = \sqrt{P_i} \frac{\mathbf{B}_i}{\|\mathbf{F}_i \mathbf{B}_i\|_F}$.
15 **end**
16 **for** $n = 1 : N_{RF_u}$ **do**
17   Calculate the phase of each entity of $\mathcal{Q}_1$, $\mathcal{V}_2$.
18   Construct the phase vectors $\boldsymbol{\psi}_n$, $\boldsymbol{\varphi}_n$.
19 **end**
20 Formulate $\boldsymbol{\Psi}_u^r$, $\boldsymbol{\Psi}_u^t$ via (3.16), (3.18).
21 Construct $\mathbf{F}_{u,r}$, $\mathbf{F}_{u,t}$ using N_{RF_u} via (3.17), (3.19).
22 **for** $z = 1 : Z$ **do**
23   Initialize the velocity as $\mathbf{Y}_z^{(0)} = \mathbf{0}$. Each entry of $\mathbf{X}_z^{(0)}$ is uniformly distributed in [0, 1].
24   Set the personal best $\mathbf{X}_{best,z}^{(0)} = \mathbf{X}_z^{(0)}$.
25 **end**
26 Find the global best $\mathbf{X}_{best}^{(0)}$ as in (3.24).
27 **for** $t = 1 : T$ **do**
28   **for** $z = 1 : Z$ **do**
29   Update the velocity $\mathbf{Y}_z^{(t)}$ as in (3.22).
30   Update the position $\mathbf{X}_z^{(t)}$ as in (3.21).
31   Find the personal best $\mathbf{X}_{best,z}^{(t)}$ as in (3.23).
32   **end**
33   Find the global best $\mathbf{X}_{best}^{(t)}$ as in (3.24).
34 **end**
35 Update $\mathbf{F}_b$, $\mathbf{F}_g$, $\mathbf{B}_b$, $\mathbf{B}_g$, $\mathbf{F}_{u,t}$, $\mathbf{F}_{u,r}$ for $\mathbf{x}$ using steps 3-21.

$$\mathbf{X}_{best}^{(t)} = \underset{\mathbf{X}_{best,i}^{(t*)}, \forall i=0,1,\cdots,Z_p}{\arg\max} \ \mathrm{R}(\mathbf{F}_1, \mathbf{B}_1, \mathbf{F}_{u,t}, \mathbf{F}_{u,r}, \mathbf{F}_2, \mathbf{B}_2, \mathbf{X}_{best,i}^{(t)}). \quad (3.24)$$

After T iterations, we assign $\mathbf{x} = \mathbf{X}_{best}^{(T)}$. Algorithm 3.1 summarizes the proposed beamforming and UAV deployment scheme.

Table 3.1 Simulation parameters

Number of antennas		$(N_T, N_t, N_r, N_R) = 144$	
Cell radius	UAV flying height	100 m	20 m
BS/gateway height	UE height	10 m	1.5–2.5 m
Path loss exponent	# paths	$\eta = 3.76$	$L = 10$
Data streams	RF chains	$N_s = 3$	$(N_{RF_b}, N_{RF_u}, N_{RF_g}) = 3$
Mean EAoD	Mean AAoD	$\theta = 60°$	$\psi = 21°$
EAoD spread	AAoD spread	$\delta^{\theta} = 15°$	$\delta^{\psi} = 11°$
Noise PSD	Channel bandwidth	− 174 dBm/Hz	10 KHz
UAV x-axis range	UAV y-axis range	$[x_{\min}, x_{\max}] = [0, 100]\, m$	$[y_{\min}, y_{\max}] = [0, 100]\, m$
Antenna spacing (in wavelength)		$d = 0.5$ m	
# network realization		1000	

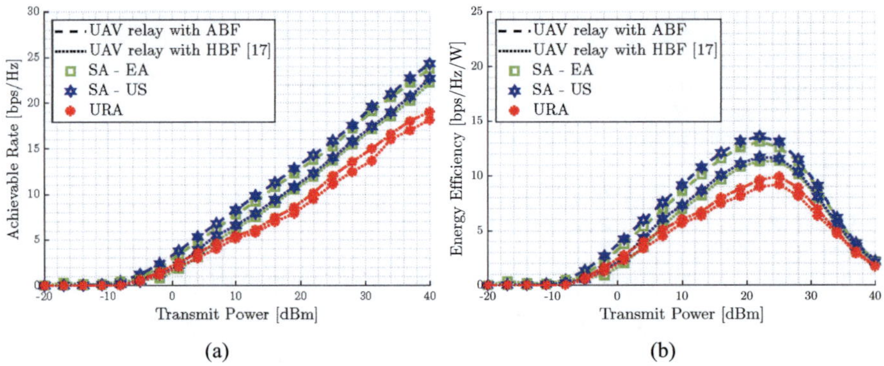

Fig. 3.4 UAV AF relaying. (**a**) AR vs. P_T. (**b**) Energy efficiency vs. P_T

3.3.4 Illustrative Results

This section presents the Monte-Carlo simulations based on the proposed scheme, with Table 3.1 outlining the simulation setup for a 3-D microcell scenario [16]. For PSO, we use $Z_p = 20$, $\gamma_1 = \gamma_2 = 2$, and $\gamma_3 = 1.1$. In Fig. 3.4a, spectral efficiency is compared for three different array configurations: URA, SA-EA, and SA-US, with AR plotted against transmit power for a fixed UAV location $(x_u, y_u) = (50, 50)$. The proposed scheme with analog beamforming at UAV relay is compared with the HBF scheme presented in [17]. The results show that both SA configurations can provide higher AR when compared to URA for both schemes, but the proposed scheme with SA-US can provide slightly better performance. For example, it can increase AR by approximately 50% when compared to URA at $P_T = 10$ dBm. The use of PSs-only RF structure at UAV relay greatly reduces the number of RF chains, and thus, the energy consumption, which is important due to limited UAV power supply. Therefore, we evaluate the energy efficiency versus transmit power for proposed AF relaying with ABF and AF relaying with HBF [17] and present the

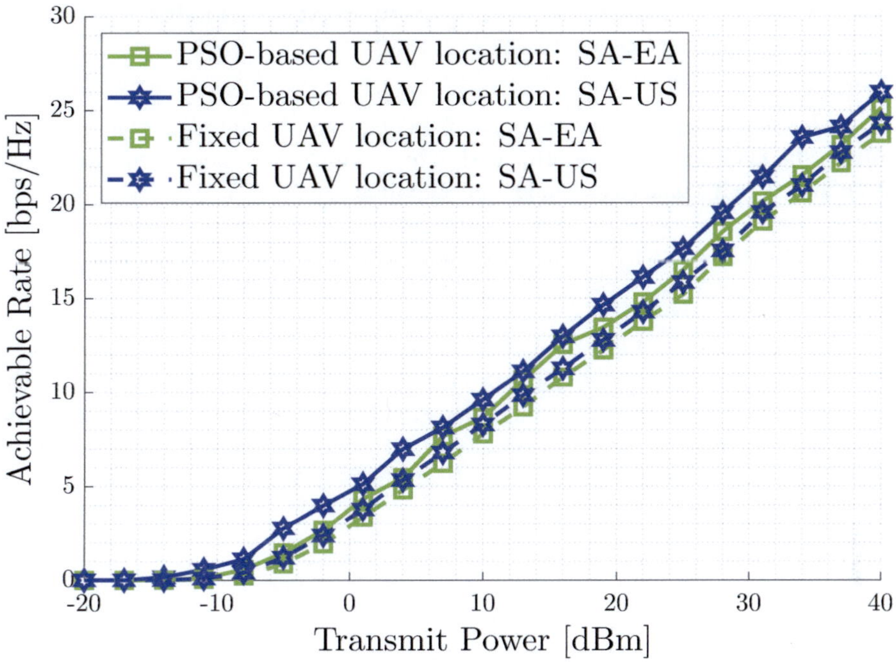

Fig. 3.5 AR vs. transmit power for different UAV locations

results in Fig. 3.4b. Here, the energy efficiency ς is defined as the ratio of AR to the total power consumption P_{total} as

$$\varsigma = \frac{R}{P_{\text{total}}} = \frac{R}{P_T + N_{RF}P_{RF} + N_{PS}P_{PS}}, \tag{3.25}$$

where P_{RF} and P_{PS} are the power consumption of each RF chain and PS, respectively. N_{PS} is the total number of PSs used at RF front end. As in [16], we use $P_T = 1$ W, $P_{RF} = 250$ mW, and $P_{PS} = 1$ mW. The results show that the proposed UAV AF relaying with RF-only structure at UAV can provide higher energy efficiency when compared to HBF in [17]. Additionally, both SA configurations provide higher ς than URA. Fig. 3.5 compares AR versus transmit power of proposed UAV AF relaying for SA configurations for two different UAV locations: (1) PSO-based UAV deployment and (2) fixed UAV location $(x_u, y_u) = (50, 50)$. It can be seen that PSO can find near-optimal UAV deployment within a given flying span $(x_{\min}, x_{\max}) = (y_{\min}, y_{\max}) = (0, 100)$, and can provide higher AR, with approximately 15% increase compared to the case when UAV is deployed at some fixed location.

3.4 Joint HBF and UAV Deployment in SU-mMIMO Systems: DF Relaying

In this section, we propose the next system model, where the problem of jointly optimizing the UAV location and HBF design to maximize the end-to-end throughput from the BS to the gateway is under consideration.

3.4.1 System Model

We consider a similar system model to the system model presented in Sect. 3.3.1, in which the UAV is the only communication link between the BS and gateway. However, this time the UAV is used as a DF relay to improve the situation as shown in Fig. 3.6. First, we define two-dimensional (2-D) and 1-dimensional (1-D) distances as follows:

$$
\begin{aligned}
\tau_{h,1} &= \sqrt{(x_b - x_u)^2 + (y_b - y_u)^2} \\
\tau_{h,2} &= \sqrt{(x_u - x_g)^2 + (y_u - y_g)^2} \\
\tau_{v,1} &= |z_b - z_u| \\
\tau_{v,2} &= \left|z_u - z_g\right|,
\end{aligned}
\tag{3.26}
$$

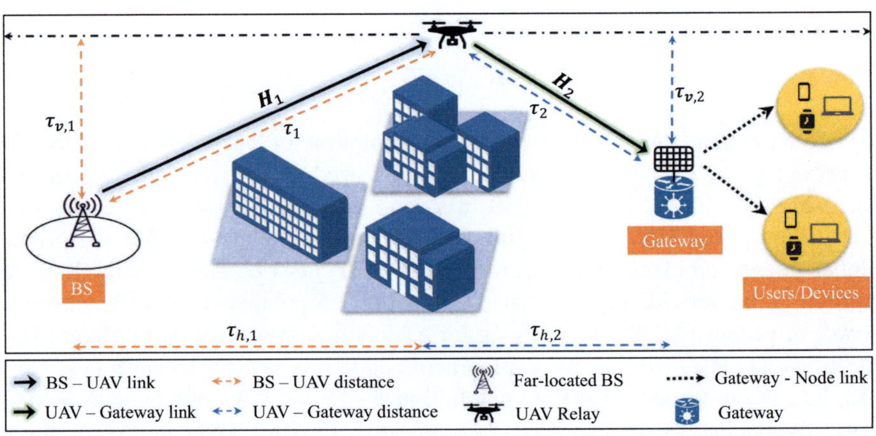

Fig. 3.6 UAV-assisted SU-mMIMO DF relaying

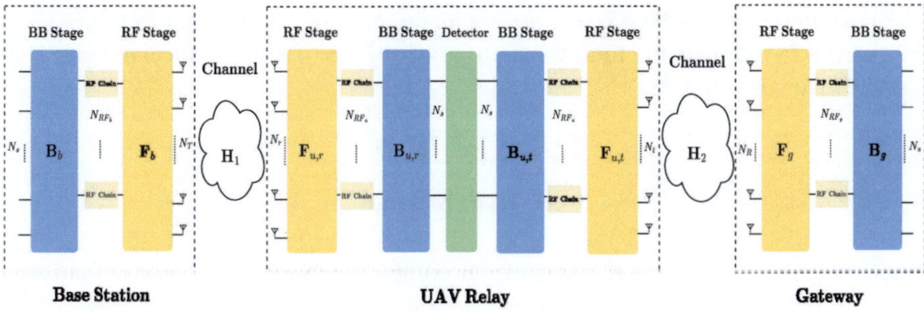

Fig. 3.7 UAV-assisted SU-mMIMO HBF DF relaying architecture

where $\tau_{h,1}$ and $\tau_{h,2}$ are the 2-D horizontal distances between UAV and BS and between UAV and gateway, respectively, and $\tau_{v,1}$ and $\tau_{v,2}$ are the 1-D vertical distances from the BS to the UAV and from the UAV to the gateway, respectively.

In the system model shown in Fig. 3.7, we consider BS is equipped with N_T antennas, UAV relay with $N_r(N_t)$ antennas for receiving (transmitting) signals, and gateway with N_R antennas. We consider HBF for all nodes (i.e., BS, UAV, and gateway), where BS consists of a RF beamforming stage $\mathbf{F}_b \in \mathbb{C}^{N_T \times N_{RF_b}}$ and BB stage $\mathbf{B}_b$ $\mathbb{C}^{N_{RF_b} \times N_s}$. Here, N_s represents the data streams from the BS, and N_{RF_b} is the RF chains such that $N_s \leq N_{RF_b} \leq N_T$ to guarantee multistream transmission. We consider half-duplex (HD) DF relaying, whereas the use of full-duplex (FD) UAV relaying is left as our future work. During the first time slot, N_s data streams are transmitted through channel $\mathbf{H}_1 \in \mathbb{C}^{N_r \times N_T}$. Using N_r antennas, UAV receives signals with RF stage $\mathbf{F}_{u,r} \in \mathbb{C}^{N_{RF_u} \times N_r}$ and BB stage $\mathbf{B}_{u,r} \in \mathbb{C}^{N_s \times N_{RF_u}}$. We assume UAV transmits the data in the second time slot using RF beamformer $\mathbf{F}_{u,t} \in \mathbb{C}^{N_t \times N_{RF_u}}$ and BB stage $\mathbf{B}_{u,t} \in \mathbb{C}^{N_{RF_u} \times N_s}$ through channel $\mathbf{H}_2 \in \mathbb{C}^{N_R \times N_t}$. The received signal at gateway is processed through RF stage $\mathbf{F}_g \in \mathbb{C}^{N_{RF_g} \times N_R}$ and BB stage $\mathbf{B}_g \in \mathbb{C}^{N_s \times N_{RF_g}}$. Here, all RF beamforming/combining stages are implemented using PSs and thus impose a CM constraint, i.e., $|\mathbf{F}_b(i,j)| = \frac{1}{\sqrt{N_T}}$, $|\mathbf{F}_{u,r}(i,j)| = \frac{1}{\sqrt{N_r}}$, $|\mathbf{F}_t(i,j)| = \frac{1}{\sqrt{N_t}}$, $|\mathbf{F}_g(i,j)| = \frac{1}{\sqrt{N_R}}$ $\forall i, j$. The design of HBF for UAV DF relaying reduces the number of RF chains from $N_r(N_t)$ to N_{RF_u}, N_T to N_{RF_b}, and N_R to N_{RF_g}, while satisfying: (1) $N_s \leq N_{RF_b} \ll N_T$; (2) $N_s \leq N_{RF_u} \ll N_r(N_t)$; and (3) $N_s \leq N_{RF_g} \ll N_R$. Considering the transmitted signal is $\mathbf{d} = [d_1, d_2, \ldots, d_{N_s}]^T \in \mathbb{C}^{N_s}$ with $\mathbb{E}\{\mathbf{d}\mathbf{d}^H\} = \mathbf{I}_{N_s} \in \mathbb{C}^{N_s \times N_s}$, then the signal transmitted by BS is given as follows:

$$\mathbf{s}_1 = \mathbf{F}_b \mathbf{B}_b \mathbf{d}. \tag{3.27}$$

The power constraint of the beamforming matrices can be expressed as $\|\mathbf{F}_b\mathbf{B}_b\|_F^2 = P_T$, where P_T denotes the total transmit power of BS. Then, the received signal at UAV is given as follows:

$$\begin{aligned}\mathbf{y}_1 &= \mathbf{H}_1\mathbf{s}_1 + \mathbf{n}_1,\\ &= \mathbf{H}_1\mathbf{F}_b\mathbf{B}_b\mathbf{d} + \mathbf{n}_1,\end{aligned} \tag{3.28}$$

where $n_1 \in \mathbb{C}^{N_r}$ denotes the zero-mean complex circularly symmetric Gaussian noise vector at UAV relay with covariance matrix $\mathbb{E}\{\mathbf{n}_1\mathbf{n}_1^H\} = \sigma_n^2\mathbf{I}_{N_r} \in \mathbb{C}^{N_r \times N_r}$. Then, the received signal after BB processing at UAV relay is written as

$$\tilde{\mathbf{y}}_1 = \mathbf{B}_{u,r}\mathbf{F}_{u,r}\mathbf{H}_1\mathbf{F}_b\mathbf{B}_b\mathbf{d} + \mathbf{B}_{u,r}\mathbf{F}_{u,r}\mathbf{n}_1. \tag{3.29}$$

UAV as DF relay uses $\tilde{\mathbf{y}}_1$ to decode the information and encodes it again for transmission in second time slot. Then, the signal transmitted by UAV is given as

$$\mathbf{s}_2 = \mathbf{F}_{u,t}\mathbf{B}_{u,t}\hat{\mathbf{d}}, \tag{3.30}$$

where $\hat{\mathbf{d}}$ is the re-encoded signal at UAV relay. After transmitting via channel $\mathbf{H}_2$, the received signal at gateway can be written as follows:

$$\tilde{\mathbf{y}_2} = \mathbf{B}_g\mathbf{F}_g\mathbf{H}_2\mathbf{F}_{u,t}\mathbf{B}_{u,t}\hat{\mathbf{d}} + \mathbf{B}_g\mathbf{F}_g\mathbf{n}_2, \tag{3.31}$$

where $\mathbf{n}_2$ denotes the additive circular symmetric Gaussian noise such that $\mathbf{n}_2 \sim \mathcal{CN}(\mathbf{0}, \sigma_n^2\mathbf{I}_{N_R})$. Then, the AR of first link (i.e., BS $\rightarrow$ UAV) is given as follows:

$$\mathrm{R}_1 = \log_2\left|\mathbf{I}_{N_s} + \mathbf{Q}_1^{-1}\mathbf{B}_{u,r}\mathcal{H}_1\mathbf{B}_b\mathbf{B}_b^H\mathcal{H}_1^H\mathbf{B}_{u,r}^H\right|, \tag{3.32}$$

where $\mathbf{Q}_1^{-1} = (\sigma_n^2\mathbf{B}_{u,r}\mathbf{F}_{u,r})^{-1}\mathbf{F}_{u,r}^H\mathbf{B}_{u,r}^H$ and $\mathcal{H}_1 = \mathbf{F}_{u,r}\mathbf{H}_1\mathbf{F}_b$. Similarly, the AR for the second link (i.e., UAV $\rightarrow$ gateway) is given as follows:

$$\mathrm{R}_2 = \log_2\left|\mathbf{I}_{N_s} + \mathbf{Q}_2^{-1}\mathbf{B}_g\mathcal{H}_2\mathbf{B}_{u,t}\mathbf{B}_{u,t}^H\mathcal{H}_2^H\mathbf{B}_g^H\right|, \tag{3.33}$$

where $\mathbf{Q}_2^{-1} = (\sigma_n^2\mathbf{B}_g\mathbf{F}_\mathbf{g})^{-1}\mathbf{F}_\mathbf{g}^H\mathbf{B}_g^H$ and $\mathcal{H}_2 = \mathbf{F}_g\mathbf{H}_2\mathbf{F}_{u,t}$. For a dual-hop mMIMO communications system, where the UAV DF relay is deployed at a fixed height z_u and relaying data to gateway, the AR can be maximized by the joint optimization of $\mathbf{F}_b$, $\mathbf{B}_b$, $\mathbf{F}_{u,t}$, $\mathbf{F}_{u,r}$, $\mathbf{B}_{u,t}$, $\mathbf{B}_{u,r}$, $\mathbf{F}_g$, $\mathbf{B}_g$, and $\tau_{h,1}$, where $\tau_{h,1}$ represents the horizontal distance between UAV and BS, which is to be optimized within the given UAV flying span. Then, we can formulate the optimization problem as

$$\begin{aligned} \max_{\{\mathbf{F}_b,\mathbf{B}_b,\mathbf{F}_{u,t},\mathbf{B}_{u,t},\mathbf{F}_{u,r},\mathbf{B}_{u,r},\mathbf{F}_g,\mathbf{B}_g,\tau_{h,1}\}} \quad & \frac{1}{2}\min(\mathrm{R}_1,\mathrm{R}_2) \\ \text{s.t.} \quad C_1: \; & |\mathbf{F}_{u,t}(i,j)| = \frac{1}{\sqrt{N_t}},\ |\mathbf{F}_{u,r}(i,j)| = \frac{1}{\sqrt{N_r}}, \quad \forall i,j, \\ C_2: \; & |\mathbf{F}_b(i,j)| = \frac{1}{\sqrt{N_T}},\ |\mathbf{F}_g(i,j)| = \frac{1}{\sqrt{N_R}} \quad \forall i,j, \\ C_3: \; & \mathbb{E}\{\|\mathbf{s}_1\|_2^2\} \le P_T,\ \mathbb{E}\{\|\mathbf{s}_2\|_2^2\} \le P_T, \\ C_4: \; & \tau_{h,\min} \le \tau_{h,1} \le \tau_{h,\max}, \end{aligned} \tag{3.34}$$

where C_1 and C_2 refer to the CM constraint due to the use of PSs for UAV, BS, and gateway, respectively, C_3 indicates the transmit power constraint for UAV and BS, and C_4 implies UAV positioning within the given flying span. Here, $\left[\tau_{h,\min}, \tau_{h,\max}\right]$ represents the UAV DF relay deployment range in x-axis and y-axis. The optimization problem defined in (3.34) is non-convex and intractable. To solve this problem, we first develop HBF stages for BS, UAV and gateway based on arbitrary UAV location to optimize $\tau_{h,1}$ using PSO. Then, based on optimal UAV relay location, we redesign RF and BB beamforming stages.

3.4.2 Joint UAV Relay Positioning and HBF

In this section, our objective is to jointly optimize the UAV relay location and HBF to reduce the CSI overhead size, while maximizing the throughput of a dual-hop UAV-assisted mMIMO communications system. First, we design the RF stages $\mathbf{F}_b, \mathbf{F}_{u,r}, \mathbf{F}_{u,t}, \mathbf{F}_g$ based on the slow time-varying AoD and AoA. Then, the BB stages $\mathbf{B}_b, \mathbf{B}_{u,r}, \mathbf{B}_{u,t}, \mathbf{B}_g$ are developed by using SVD.

3.4.2.1 RF-Beamformers Design

We first define AoD and AoA supports as

$$\mathrm{AoD} = \left\{\sin(\theta)\left[\cos(\psi), \sin(\psi)\right] \middle| \theta \in \boldsymbol{\theta}_t, \psi \in \boldsymbol{\psi}_t\right\}, \tag{3.35}$$

$$\mathrm{AoA} = \left\{\sin(\theta)\left[\cos(\psi), \sin(\psi)\right] \middle| \theta \in \boldsymbol{\theta}_r, \psi \in \boldsymbol{\psi}_r\right\}, \tag{3.36}$$

where $\boldsymbol{\theta}_i = \left[\theta_i - \delta_i^\theta, \theta_i + \delta_i^\theta\right]$ and $\boldsymbol{\psi}_i = \left[\psi_i - \delta_i^\psi, \psi_i + \delta_i^\psi\right]$ denote the azimuth and elevation angle supports, respectively. To exploit all DoF provided by the channel to maximize the transmit (receive) beamforming gain in either channel link, we consider selecting the columns of $\mathbf{F}_b(\mathbf{F}_{u,r})$ and $\mathbf{F}_{u,t}(\mathbf{F}_g)$ from the subspace spanned by $\boldsymbol{\phi}_t (\boldsymbol{\phi}_r)$ in first and second links, respectively (i.e., Span $\{\mathbf{F}_b, \mathbf{F}_{u,t}\} \subset$ Span $(\boldsymbol{\phi}_t)$ and Span $\{\mathbf{F}_{u,r}, \mathbf{F}_g\} \subset$ Span $(\boldsymbol{\phi}_r)$). Then, the transmit RF beamformers are constructed via transmit steering vector as

$$\begin{aligned}\mathbf{e}_t\,(\theta,\psi) = \frac{1}{\mathcal{M}_t}&\left[1, e^{j2\pi d \sin(\theta)\cos(\psi)}, \cdots, e^{j2\pi d(\mathcal{M}_{x,t}-1)\sin(\theta)\cos(\psi)}\right]^T \\ &\otimes\left[1, e^{j2\pi d \sin(\theta)\sin(\psi)}, \cdots, e^{j2\pi d(\mathcal{M}_{y,t}-1)\sin(\theta)\sin(\psi)}\right]^T,\end{aligned} \tag{3.37}$$

where $\mathcal{M}_t = \{N_T, N_t\}$. We define N orthogonal steering vectors via generating the quantized angle pairs as $\lambda^u_{x,t} = -1 + \frac{2u-1}{\mathcal{M}_{x,t}}$ for $u = 1, \cdots, \mathcal{M}_{x,t}$ and $\lambda^k_{y,t} = -1 + \frac{2k-1}{\mathcal{M}_{y,t}}$ for $k = 1, \cdots, \mathcal{M}_{y,t}$. The use of quantized angle pairs minimizes the RF chain utilization, while covering the complete AoD and AoA supports. Then, the quantized angle pairs inside the AoD support satisfying (3.35) are obtained as

$$(\lambda^u_{x,t}, \lambda^k_{y,t}) \mid \sin(\theta)\cos(\psi) \in \boldsymbol{\lambda}^u_{x,t}, \sin(\theta)\sin(\psi) \in \boldsymbol{\lambda}^k_{y,t}, \tag{3.38}$$

where $\boldsymbol{\lambda}^u_{x,t} = \left[\lambda^u_{x,t} - \frac{1}{\mathcal{M}_{x,t}}, \lambda^u_{x,t} + \frac{1}{\mathcal{M}_{x,t}}\right]$ and $\boldsymbol{\lambda}^k_{y,t} = [\lambda^k_{y,t} - \frac{1}{\mathcal{M}_{y,t}}, \lambda^k_{y,t} + \frac{1}{\mathcal{M}_{y,t}}]$ are the boundaries of $\lambda^u_{x,t}$ and $\lambda^k_{y,t}$, respectively. By using (3.35) and (3.38), the transmit RF beamformers for BS and UAV are found as

$$\mathbf{F}_b = \left[\mathbf{e}_t\left(\lambda^{u_1}_{x,t}, \lambda^{k_1}_{y,t}\right), \ldots, \mathbf{e}_t\left(\lambda^{u_{N_{RF_b}}}_{x,t}, \lambda^{k_{N_{RF_b}}}_{y,t}\right)\right] \in \mathbb{C}^{N_T \times N_{RF_b}} \tag{3.39}$$

$$\mathbf{F}_{u,t} = \left[\mathbf{e}_t\left(\lambda^{u_1}_{x,t}, \lambda^{k_1}_{y,t}\right), \ldots, \mathbf{e}_t\left(\lambda^{u_{N_{RF_u}}}_{x,t}, \lambda^{k_{N_{RF_u}}}_{y,t}\right)\right] \in \mathbb{C}^{N_t \times N_{RF_u}}. \tag{3.40}$$

Following similar approach, the receive RF beamformers can be designed via receive steering vector as

$$\begin{aligned}\mathbf{e}_r\,(\theta,\psi) = \frac{1}{\mathcal{M}_r}&\left[1, e^{-j2\pi d \sin(\theta)\cos(\psi)}, \cdots, e^{-j2\pi d(\mathcal{M}_{x,r}-1)\sin(\theta)\cos(\psi)}\right]^T \\ &\otimes\left[1, e^{-j2\pi d \sin(\theta)\sin(\psi)}, \cdots, e^{-j2\pi d(\mathcal{M}_{y,r}-1)\sin(\theta)\sin(\psi)}\right]^T,\end{aligned} \tag{3.41}$$

where $\mathcal{M}_r = \{N_R, N_r\}$. Using quantized angle pairs as $\lambda^u_{x,r} = -1 + \frac{2u-1}{\mathcal{M}_{x,r}}$ for $u = 1, \cdots, \mathcal{M}_{x,r}$ and $\lambda^k_{y,r} = -1 + \frac{2k-1}{\mathcal{M}_{y,r}}$ for $k = 1, \cdots, \mathcal{M}_{y,r}$, we can formulate the receive RF beamformers for the UAV and gateway as

$$\mathbf{F}_{u,r} = \left[\mathbf{e}_r\left(\lambda^{u_1}_{x,r}, \lambda^{k_1}_{y,r}\right), \ldots, \mathbf{e}_r\left(\lambda^{u_{N_{RF_u}}}_{x,r}, \lambda^{k_{N_{RF_u}}}_{y,r}\right)\right]^T \in \mathbb{C}^{N_{RF_u} \times N_r} \tag{3.42}$$

$$\mathbf{F}_g = \left[\mathbf{e}_r\left(\lambda^{u_1}_{x,r}, \lambda^{k_1}_{y,r}\right), \ldots, \mathbf{e}_r\left(\lambda^{u_{N_{RF_g}}}_{x,r}, \lambda^{k_{N_{RF_g}}}_{y,r}\right)\right]^T \in \mathbb{C}^{N_R \times N_{RF_g}}, \tag{3.43}$$

Algorithm 3.2: Proposed joint HBF and UAV positioning and HBF algorithm

1 **Input**: Z_p, T, (θ, ψ), (x_b, y_b, z_b), (x_g, y_g, z_g), (x_u, y_u, z_u).
2 **Output**: $\tau_{h,1}$, $\mathbf{F}_b$, $\mathbf{B}_b$, $\mathbf{F}_{u,t}$, $\mathbf{B}_{u,t}$, $\mathbf{F}_{u,r}$, $\mathbf{B}_{u,r}$, $\mathbf{F}_g$, $\mathbf{B}_g$.
3 Formulate transmit RF beamformers using (3.39), (3.40).
4 Formulate receive RF beamformers using (3.42), (3.43).
5 Using (3.45), (3.46), form the unitary matrices $\mathcal{Q}_i$ and $\mathcal{V}_i$.
6 Construct BB stages via (3.47), (3.48).
7 **for** $i = 1 : Z_p$ **do**
8   Initialize the velocity as $V_i^{(0)} = 0$.
9   Each entry of $\tau_i^{(0)}$ is uniformly distributed in [0, 1].
10   Set the personal best $\tau_{best,i}^{(0)} = \tau_i^{(0)}$.
11 **end**
12 Find the global best $\tau_{best}^{(0)}$ as in (3.53).
13 **for** $t = 1 : T$ **do**
14   **for** $i = 1 : Z_p$ **do**
15   Update the velocity $V_i^{(t)}$ as in (3.50).
16   Update the position $\tau_i^{(t)}$ as in (3.51).
17   Find the personal best $\tau_{best,i}^{(t)}$ as in (3.52).
18   **end**
19   Find the global best $\tau_{best}^{(t)}$ as in (3.53).
20 **end**
21 Update $\mathbf{B}_b$, $\mathbf{B}_g$, $\mathbf{B}_{u,t}$, $\mathbf{B}_{u,r}$ for $\tau_{h,1}$ using steps 1-4.

where the quantized angle pairs inside the AoA support satisfying (3.36) are obtained as

$$(\lambda_{x,r}^u, \lambda_{y,r}^k) \mid \sin(\theta)\cos(\psi) \in \boldsymbol{\lambda}_{x,r}^u, \sin(\theta)\sin(\psi) \in \boldsymbol{\lambda}_{y,r}^k, \tag{3.44}$$

where $\boldsymbol{\lambda}_{x,r}^u = [\lambda_{x,r}^u - \frac{1}{M_{x,r}}, \lambda_{x,r}^u + \frac{1}{M_{x,r}}]$ is the boundary of $\lambda_{x,r}^u$, and $\boldsymbol{\lambda}_{y,r}^k = [\lambda_{y,r}^k - \frac{1}{M_{y,r}}, \lambda_{y,r}^k + \frac{1}{M_{y,r}}]$ is the boundary of $\lambda_{y,r}^k$.

3.4.2.2 BB Stage Design and UAV Deployment

After designing the transmit and receive RF beamformers for BS, UAV, and gateway, the effective channel matrices for first and second links as seen from the BB stages are given as

$$\mathcal{H}_1 = \mathbf{F}_{u,r}\mathbf{H}_1\mathbf{F}_b \tag{3.45}$$

$$\mathcal{H}_2 = \mathbf{F}_g\mathbf{H}_2\mathbf{F}_{u,t}. \tag{3.46}$$

By using SVD of effective channel matrix as $\mathcal{H}_i = \mathcal{Q}_i \boldsymbol{\Sigma}_i \mathcal{V}_i^H$, where $\mathcal{Q}_i \in \mathbb{C}^{N_{\mathcal{RF}_\mathcal{R}} \times rank(\mathcal{H}_i)}$ and $\mathcal{V}_i \in \mathbb{C}^{N_{\mathcal{RF}_\mathcal{T}} \times rank(\mathcal{H}_i)}$ are tall unitary matrices and $\boldsymbol{\Sigma}_i$ is the diagonal matrix with singular values in the decreasing order. Here, $\mathcal{N}_{\mathcal{RF}_\mathcal{T}} = \{N_{RF_b}, N_{RF_u}\}$ and $\mathcal{N}_{\mathcal{RF}_\mathcal{R}} = \{N_{RF_g}, N_{RF_u}\}$ represent the transmit and receive RF chains, respectively. Assuming $\text{rank}(\mathcal{H}_i) \geq N_s$, $\mathcal{V}_i$ can be partitioned as $\mathcal{V}_i = [\mathcal{V}_{i,1}, \mathcal{V}_{i,2}]$ with $\mathbf{V}_{i,1} \in \mathbb{C}^{\mathcal{N}_{\mathcal{RF}_\mathcal{T}} \times N_s}$. Then, the optimal $\mathbf{B}_b$ and $\mathbf{B}_{u,r}$ in first link can be obtained as [19]

$$\mathbf{B}_b = \sqrt{\frac{P_T}{N_s}} \mathcal{V}_1, \qquad \mathbf{B}_{u,r} = \mathcal{Q}_1^H. \tag{3.47}$$

Similarly, the optimal $\mathbf{B}_g$ and $\mathbf{B}_{u,t}$ in second link can be obtained as

$$\mathbf{B}_{u,t} = \mathcal{Q}_2^H, \qquad \mathbf{B}_g = \sqrt{\frac{P_T}{N_s}} \mathcal{V}_2. \tag{3.48}$$

The design of RF and BB stages reduces the optimization problem given in (3.34) by satisfying the constraints $C_1 - C_3$. Then, the optimization problem can be reformulated as

$$\begin{aligned} \max_{\tau_{h,1}} \quad & \frac{1}{2} \min\left(\text{R}_1, \text{R}_2\right) \\ \text{s.t.} \quad & C_4 : \tau_{h,\min} \leq \tau_{h,1} \leq \tau_{h,\max}. \end{aligned} \tag{3.49}$$

This resulting problem in (3.49) is still NP hard as the UAV location affects the CSI, which, in turn, affects the beamforming design of the RF and BB stages. Therefore, the problem is combinatorial and involves a large search space, making it difficult to find the optimal solution within a reasonable time frame. Hence, we propose a PSO-based algorithmic solution for jointly optimizing UAV location and sequentially updating both RF and BB stages for BS, UAV, and gateway. The algorithm starts with a swarm of Z_p particles, each with its own position, velocity, and fitness value, which are randomly placed in search space. Let $\tau_i^{(t)}$ and $V_i^{(t)}$ be the position and velocity of ith particle during tth iteration, respectively. Then, during the iterations, both $\tau_i^{(t)}$ and $V_i^{(t)}$ of ith particles are updated as follows:

$$V_i^{(t+1)} = \gamma_1 z_1 (\tau_{best}^{(t)} - \tau_i^{(t)}) + \gamma_2 z_2 (\tau_{best,i}^{(t)} - \tau_i^{(t)}) + \gamma_3 V_i^{(t)}, \tag{3.50}$$

$$\tau_i^{(t+1)} = \tau_i^{(t)} + V_i^{(t+1)}, \tag{3.51}$$

where γ_1, γ_2, and γ_3 are tuning parameters, and z_1 and z_2 are random numbers between 0 and 1. Each ith particle represents the candidate position of UAV, and it iteratively updates its personal best and global best solution during a total of T

iterations as

$$\tau_{best,i}^{(t)} = \underset{\tau_i^{(t^*)}, \forall t^*=0,1,\cdots,t}{\arg\max} \frac{1}{2} \min\left(\mathrm{R}_1(\tau_i^{(t^*)}), \mathrm{R}_2(\tau_i^{(t^*)})\right), \tag{3.52}$$

$$\tau_{best}^{(t)} = \underset{\tau_{best,i}^{(t^*)}, \forall i=0,1,\cdots,Z_p}{\arg\max} \frac{1}{2} \min\left(\mathrm{R}_1(\tau_{best,i}^{(t)}), \mathrm{R}_2(\tau_{best,i}^{(t)})\right). \tag{3.53}$$

After T iterations, we assign $\tau_{h,1} = \tau_{h,1,best}^{(T)}$. Algorithm 3.2 summarizes the proposed joint HBF and UAV relay deployment scheme. The developed algorithm is applicable to both fixed and movable nodes.

3.4.3 Illustrative Results

In this section, the Monte Carlo simulation results are presented based on the proposed scheme. Table 3.2 outlines the simulation setup based on the 3-D microcell scenario [16]. We consider $Z_p = 20, \gamma_1 = \gamma_2 = 2$, and $\gamma_3 = 1.1$ for PSO parameters. In Fig. 3.8a, we plot the AR versus transmit power for a fixed UAV location $\left(\tau_{h,1} = \tau_{h,2} = 50\text{m}\right)$ for both FDBF and HBF for $N_s = \{1, 2, 3\}$. FDBF provides higher performance than HBF at the expense of an increased number of RF chains (i.e., $N_{RF} = 64$), whereas HBF can provide performance close to FDBF by only using $N_{RF} = 4$ (i.e., reduction of 93.75% in RF chains). Moreover, by increasing the data streams N_s, AR can be significantly increased (e.g., by

Table 3.2 Simulation parameters

Number of antennas		$(N_T, N_t, N_r, N_R) = 64$	
Number of paths	Path loss exponent	$L = 10$	$\eta = 3.6$
Frequency	Channel bandwidth	28 GHz	100 MHz
Noise PSD	Reference path loss α	− 174 dBm/Hz	61.34 dB
BS/Gateway height	UAV height	10 m	20 m
Horizontal distance τ_h	Vertical distance $\tau_{v,1}$	100 m	100 m
UAV x-axis range	UAV y-axis range	$[x_{\min}, x_{\max}] =$ [0, 100]m	$[y_{\min}, y_{\max}] =$ [0, 100]m
Azimuth AoD/AoA (1st link)	Azimuth AoD/AoA (2nd link)	120°	150°
Elevation AoD/AoA (1st link)	Elevation AoD/AoA (2nd link)	60°	30°
Azimuth/elevation angle spread	# of network realizations	± 10°	2000
Minimum horizontal distance $\tau_{h,\min}$	Maximum horizontal distance $\tau_{h,\max}$	0	100 m

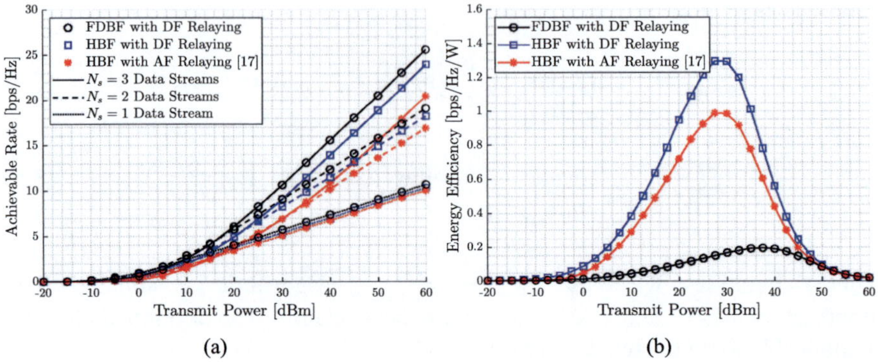

Fig. 3.8 Proposed DF HBF vs. FDBF and AF relaying: (**a**) AR comparison. (**b**) Energy efficiency comparison

increasing N_s from 1 to 3, AR can increase from 6 bps/Hz to 13 bps/Hz for HBF at $P_T = 40$ dBm, which represents AR increase of around 100%). We also compare AR of the proposed HBF for DF relaying with AF relaying [17]. It can be seen that the proposed HBF with DF relaying can achieve higher performance than AF (e.g., DF can provide a gain of around 40% at $P_T = 40$ dBm). In Fig. 3.8b, we plot the energy efficiency versus transmit power for FDBF with DF relaying and HBF with DF and AF relaying.

The energy efficiency ς_i is defined similar to (3.25) as

$$\varsigma_i = \frac{R_i}{P_{\text{total}}} = \frac{R_i}{P_T + N_{RF}P_{RF} + N_{PS}P_{PS}}, \tag{3.54}$$

where $i = \{DF, AF\}$. Likewise, we use $P_T = 1$ W, $P_{RF} = 250$ mW, and $P_{PS} = 1$ mW. FDBF with DF relaying achieves very low ς due to the use of a large number of RF chains. The proposed HBF with DF relaying can provide a higher energy efficiency for both FDBF and HBF with AF relaying.

Figure 3.9 compares AR versus transmit power of proposed HBF with DF relaying using $N_s = 3$ for different fixed UAV locations with $\tau_{h,1} \in \{0, 10, 20, 30, 40, 50\}$ m. The results show that the UAV deployed close to the BS or gateway can achieve a lower AR; however, the performance can be improved by deploying the UAV relay around halfway between BS and gateway. However, by using the proposed PSO-based deployment scheme, we can achieve a near-optimal UAV location, which can provide a higher rate. Finally, Fig. 3.10 shows the distribution of realizations out of 2000 total network realizations for best UAV relay location at different $\tau_{h,1}$, where $\tau_{h,1}$ varies between 0 and 100 m. The distribution shows a higher proportion of realizations for the UAV best location within [40, 60] m (i.e., around 80% of the total realizations). More specifically, around 50% of the distribution lies within [45, 55] m, which shows the optimal deployment for a UAV relay in dual-hop communications is in the midst of the BS and gateway.

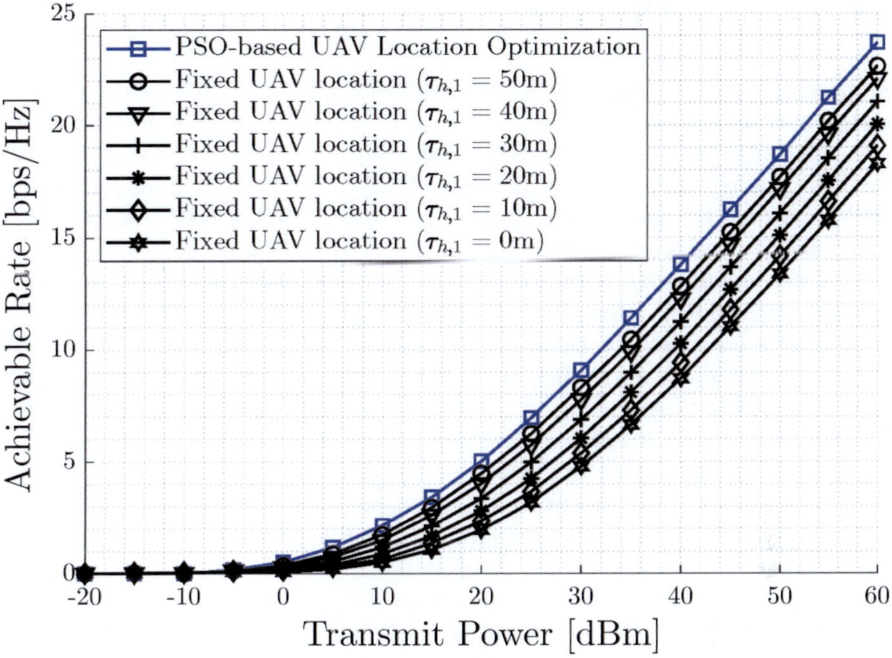

Fig. 3.9 AR vs. transmit power for the proposed PSO-based HBF

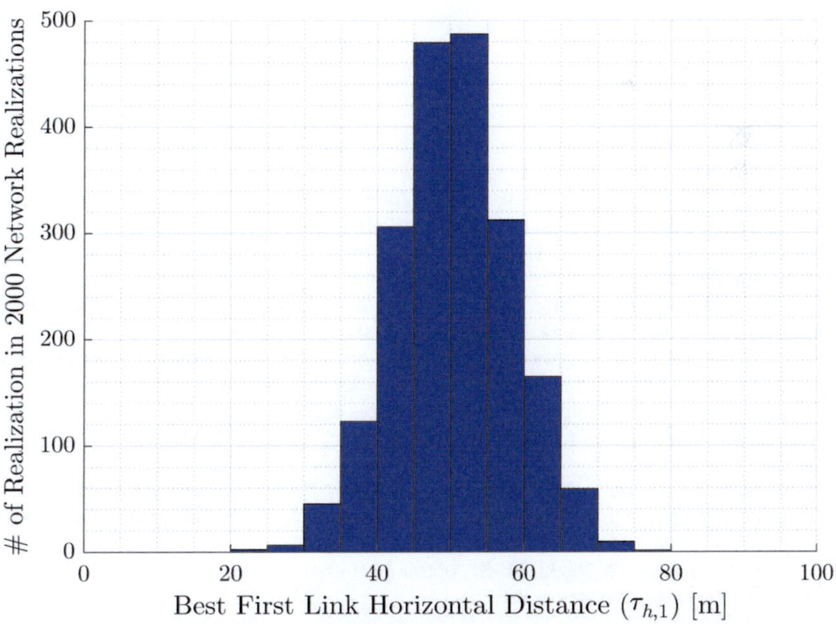

Fig. 3.10 Distribution for the best UAV location using $N_s = 3$, $P_T = 40$ dBm

3.5 Concluding Remarks

In this chapter, we investigated two UAV-assisted dual-hop SU-mMIMO communication systems. We presented scenarios that involve BS-gateway communications, where the only mean of communication is through a UAV relay. First, we considered the UAV acting as an AF relay and addressed the problem of jointly designing HBF stages for BS and gateway (using OMP-based solution), ABF for the UAV relay, and its deployment within a given flying span. We propose a PSO-based algorithmic solution that optimizes the UAV location and sequentially design RF and BB stages for all three nodes, leading to higher ARs. The illustrative results show that the proposed joint beamforming and UAV positioning scheme can provide higher spectral and energy efficiencies compared to existing HBF schemes. Then we considered the UAV acting as a DF relay with HBF setup for all three nodes. The PSO-based UAV location optimization is considered, where the RF stages are designed using the angular location of nodes, reducing the number of RF chains, and BB stages are designed using reduced-dimension effective channel utilizing SVD. The results illustrate that the proposed HBF approach leads to significantly improved energy efficiency when compared to FDBF and can find optimal UAV deployment for enhanced coverage and capacity.

Although the ABF achieved superior results than HBF in SU-mMIMO, but it has limitation in MU-mMIMO scenarios due to the phase-only structure. Therefore, we explore HBF solutions for MU-mMIMO cases, which has shown a significantly higher energy efficiency than FDBF, while having a satisfactory performance, in the following chapters.

References

1. S. Zhang, H. Zhang, Q. He, K. Bian, L. Song, Joint trajectory and power optimization for UAV relay networks. IEEE Commun. Lett. **22**(1), 161–164 (2018)
2. Y. Chen, W. Feng, G. Zheng, Optimum placement of UAV as relays. IEEE Commun. Lett. **22**(2), 248–251 (2017)
3. X. Jiang, Z. Wu, Z. Yin, Z. Yang, Power and trajectory optimization for UAV-enabled Amplify-and-Forward relay networks. IEEE Access **6**, 48688–48696 (2018)
4. F. Ono, H. Ochiai, R. Miura, A wireless relay network based on unmanned aircraft system with rate optimization. IEEE Trans. Wireless Commun. **15**(11), 7699–7708 (2016)
5. Y. Guo, S. Yin, J. Hao, Joint placement and resources optimization for multi-user UAV-relaying systems with underlaid cellular networks. IEEE Trans. Veh. Technol. **69**(10), 12374–12377 (2020)
6. W. Wang, N. Qi, L. Jia, C. Li, T.A. Tsiftsis, M. Wang, Energy-efficient UAV-relaying 5G/6G spectrum sharing networks: Interference coordination with power management and trajectory design. IEEE Open J. Commun. Soc. **3**, 1672–1687 (2022)
7. L. Sboui, H. Ghazzai, Z. Rezki, M.-S. Alouini, Achievable rates of UAV-relayed cooperative cognitive radio MIMO systems. IEEE Access **5**, 5190–5204 (2017)
8. L. Zhu, J. Zhang, Z. Xiao, X.-G. Xia, R. Zhang, Multi-UAV aided millimeter-wave networks: positioning, clustering, and beamforming. IEEE Trans. Wireless Commun. **21**(7), 4637–4653 (2022)

9. J. Du, W. Xu, Y. Deng, A. Nallanathan, L. Vandendorpe, Energy-saving UAV-assisted multiuser communications with massive MIMO hybrid beamforming. IEEE Commun. Lett. **24**(5), 1100–1104 (2020)
10. X. Pang, J. Tang, N. Zhao, X. Zhang, Y. Qian, Energy-efficient design for mmWave-enabled NOMA-UAV networks. Sci. China Inf. Sci. **64**, 1–14 (2021)
11. W. Feng et al., Hybrid beamforming design and resource allocation for UAV-aided wireless-powered mobile edge computing networks with NOMA. IEEE J. Sel. Areas Commun. **39**(11), 3271–3286 (2021)
12. K. Liu, Y. Liu, P. Yi, Z. Xiao, X.-G. Xia, Deployment and robust hybrid beamforming for UAV MmWave communications. IEEE Trans. Commun. **71**(5), 3073–3086 (2023)
13. X. Yu, X. Huang, K. Wang, F. Shu, X. Dang, Joint design of power allocation, beamforming, and positioning for energy-efficient UAV-aided multiuser millimeter-wave systems. IEEE J. Sel. Areas Commun. **40**(10), 2930–2945 (2022)
14. W. Feng et al., Joint 3D trajectory and power optimization for UAV-aided mmWave MIMO-NOMA networks. IEEE Trans. Commun. **69**(4), 2346–2358 (2021)
15. R. Méndez-Rial, C. Rusu, N. González-Prelcic, A. Alkhateeb, R.W. Heath, Hybrid MIMO architectures for millimeter wave communications: phase shifters or switches? IEEE Access **4**, 247–267 (2016)
16. A. Koc, A. Masmoudi, T. Le-Ngoc, 3D angular-based hybrid precoding and user grouping for uniform rectangular arrays in massive MU-MIMO systems. IEEE Access **8**, 84689–84712 (2020)
17. Y. Zhang, M. Xiao, S. Han, M. Skoglund, W. Meng, On precoding and energy efficiency of full-duplex millimeter-wave relays. IEEE Trans. Wirel. Commun. **18**(3), 1943–1956 (2019)
18. A. Koc, T. Le-Ngoc, Swarm intelligence based power allocation in hybrid millimeter-wave massive MIMO systems, in *Proceedings of the IEEE Wireless Communications and Networking Conference (WCNC)* (2021), pp. 1–7
19. A. Koc, T. Le-Ngoc, Hybrid millimeter-wave massive MIMO systems with low CSI overhead and few-bit DACs/ADCs, in *Proceedings of the IEEE 92th Vehicle Technology Conference (VTC-Fall)*, (2020), pp. 1–5

Chapter 4
PSO-Based UAV-Assisted DF Relaying in Terrestrial MU-mMIMO System

4.1 Introduction

In this chapter, we will extend the problem to MU-mMIMO settings where the UAV operates as a DF relay using HBF architecture. Specifically, the UAV directly forwards data from the BS to multiple ground users/devices rather than forwarding data to a gateway, making an MU communication scenario. Furthermore, we also consider the MU PA jointly with UAV location optimization to further improve the performance.

In some of the previous research including UAVs, the problems are limited to SU-mMIMO or single-antenna scenarios [1–5], while we observed several research that overlooked the relying aspects (e.g., [6, 7]) and beamforming solutions (e.g., [8, 9]) in the previous chapter. The joint optimization of UAV location, PA, and HBF design for a dual-hop mmWave MU-mMIMO communications networks is an unaddressed problem, presenting a significant opportunity to advance the field of UAV-based wireless communications. To address this gap in literature, this chapter aims to highlight the full potential of beamforming, and its ability to enhance performance in dual-hop MU-mMIMO UAV communications networks. The main contributions of this work are summarized as follows:

- We propose three novel optimization schemes for maximizing total capacity in UAV-assisted MU-mMIMO communications systems: (1) J-HBF-PSOPA-FL; (2) J-HBF-PSOL-EQPA; and (3) J-HBF-PSOLPA. Using swarm intelligence-based PSO, we tackle the challenging non-convex problem with high-dimensional variable matrices and fractional programming variables, while adhering to constraints like UAV deployment span, total transmit power, PA, and CM. In particular, the RF stages are formulated using slow time-varying AoD/AoA information, while the BB stages use the reduced-dimension effective channel information with RZF. In J-HBF-PSOPA-FL, we allocate optimal power for multiple users with fixed UAV location and design the RF and BB stages for

T. Le-Ngoc et al., *AI-Enabled UAV-Assisted Massive MIMO*, Wireless Networks,
https://doi.org/10.1007/978-3-032-07872-8_4

maximum capacity. In J-HBF-PSOL-EQPA, we optimize UAV location for equal PA, whereas, in J-HBF-PSOLPA, we jointly optimize both UAV location and PA and design HBF stages for both UAV and BS. The illustrative results show that J-HBF-PSOLPA can achieve higher total rate compared to J-HBF-PSOL-EQPA and J-HBF-PSOPA-FL schemes.

- We analyze the performance of UAV DF relaying with and without buffering. Most existing studies have explored the option of forwarding the received signal without buffering the data, which fails to fully leverage the UAV's mobility [1–3, 6–17]. The proposed solutions incorporate signal buffering at the UAV, enhancing the relaying performance significantly. Particularly, we analyze the average delay of delay-unconstrained and delay-constrained transmissions, which can be significantly reduced by J-HBF-PSOLPA compared to joint HBF, fixed UAV location and equal PA (J-HBF-FL-EQPA).

The rest of this chapter is organized as follows. Section 4.2 discusses the system model and channel model of mmWave UAV-assisted MU-mMIMO systems followed by the proposed joint HBF, PA, and UAV location optimization schemes in Sect. 4.3. The illustrative results are presented in Sect. 4.4 followed by the conclusions in Sect. 4.5.

4.2 System and Channel Model

This section presents the system and channel model of the proposed UAV-assisted mmWave MU-mMIMO communications systems.

4.2.1 System Model

In this chapter, we explore a more challenging scenario, where multiple users/devices are directly connected to the UAV rather than the gateway, establishing a multiuser setup. The users are situated in a remote area that is difficult to access directly by BS/eNodeB due to various obstacles such as buildings, mountains, etc. In particular, the UAV is introduced as a dual-hop DF relay to access the users, as depicted in Fig. 4.1. Similar to previous chapter, we assume that there is no direct link between BS and users due to severe blockage. Let (x_b, y_b, z_b), (x_u, y_u, z_u), and (x_k, y_k, z_k) represent the coordinates of the BS, UAV relay, and

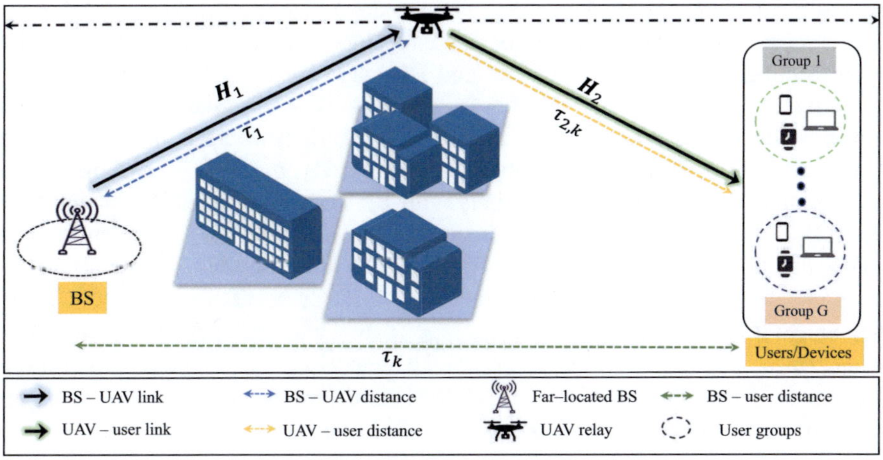

Fig. 4.1 UAV-assisted relaying in MU-mMIMO communications systems

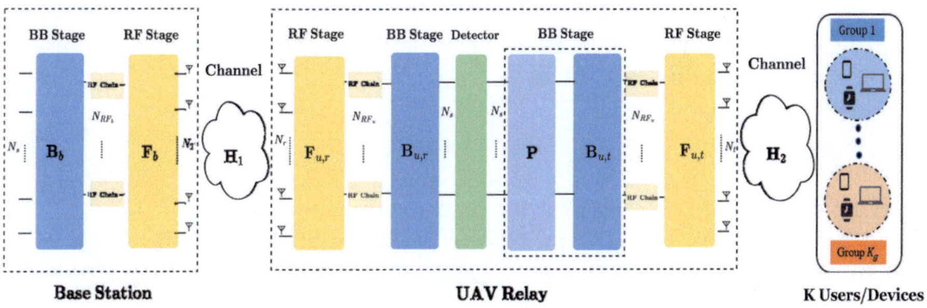

Fig. 4.2 System model of UAV-assisted DF relaying in MU-mMIMO networks

the kth user, respectively. We define the 3-D distances for a UAV-assisted mmWave MU-mMIMO system in the following manner:

$$\begin{aligned}\tau_1 &= \sqrt{(x_u - x_b)^2 + (y_u - y_b)^2 + (z_u - z_b)^2} \\ \tau_{2,k} &= \sqrt{(x_u - x_k)^2 + (y_u - y_k)^2 + (z_u - z_k)^2} \\ \tau_k &= \sqrt{(x_b - x_k)^2 + (y_b - y_k)^2 + (z_b - z_k)^2},\end{aligned} \tag{4.1}$$

where τ_1, $\tau_{2,k}$, and τ_k denote the 3-D distance between UAV and BS, between the UAV and kth user, and between BS and kth user, respectively.

In the system model depicted in Fig. 4.2, we consider BS equipped with N_T antennas, UAV relay with N_r antennas for receiving, and N_t antennas for sending information to K single-antenna users scattered in G groups, where gth group has

K_g users such that $K = \sum_{g=1}^{G} K_g$. For the downlink transmission of $N_s = K$ data streams, we consider HBF for BS and UAV, where BS consists of an RF beamforming stage $\mathbf{F}_b \in \mathbb{C}^{N_T \times N_{RF_b}}$ and BB stage $\mathbf{B}_b \in \mathbb{C}^{N_{RF_b} \times K}$, where N_{RF_b} is the number of RF chains such that $N_s \leq N_{RF_b} \leq N_T$ to guarantee multistream transmission. We assume half-duplex (HD) DF relaying, where K data streams are transmitted from BS by N_T antennas via channel $\mathbf{H}_1 \in \mathbb{C}^{N_r \times N_T}$, and then using N_r antennas, UAV receives the signals with RF stage $\mathbf{F}_{u,r} \in \mathbb{C}^{N_{RF_u} \times N_r}$ and BB stage $\mathbf{B}_{u,r} \in \mathbb{C}^{K \times N_{RF_u}}$ during the first time slot , where N_{RF_u} is the number of RF chains at UAV transmit/receive. Afterward, within the second time slot, UAV transmits the data using RF beamformer $\mathbf{F}_{u,t} = [\mathbf{f}_{u,t,1} \cdots, \mathbf{f}_{u,t,N_{RF_u}}] \in \mathbb{C}^{N_t \times N_{RF_u}}$, BB stage $\mathbf{B}_{u,t} = [\mathbf{b}_{u,t,1}, \cdots, \mathbf{b}_{u,t,K}] \in \mathbb{C}^{N_{RF_u} \times K}$, and MU PA matrix $\mathbf{P} = \text{diag}(\sqrt{p_1}, \cdots, \sqrt{p_K}) \in \mathbb{C}^{K \times K}$ through channel $\mathbf{H}_2 \in \mathbb{C}^{K \times N_t}$, where p_k reflects the allocated power to kth user. Here, all RF beamforming and combining stages are implemented using PSs and thus impose a CM constraint, i.e., $|\mathbf{F}_b(i, j)| = \frac{1}{\sqrt{N_T}}$, $|\mathbf{F}_{u,r}(i, j)| = \frac{1}{\sqrt{N_r}}$, $|\mathbf{F}_{u,t}(i, j)| = \frac{1}{\sqrt{N_t}} \forall i, j$. The HBF design substantially reduces the number of RF chains, decreasing them from N_T to N_{RF_b} for the BS and from N_t (N_r) to N_{RF_u} for the UAV, while meeting the following conditions: (1) $K \leq N_{RF_b} \ll N_T$; and (2) $K \leq N_{RF_u} \ll N_r$ (N_t). Considering K data streams, the signal transmitted by BS is given as

$$\mathbf{s}_1 = \mathbf{F}_b \mathbf{B}_b \mathbf{d}. \tag{4.2}$$

Here, $\mathbf{d} = [d_1, d_2, \cdots, d_K]^T$ is the data stream with $\mathbb{E}\{\mathbf{d}\mathbf{d}^H\} = \mathbf{I}_K \in \mathbb{C}^{K \times K}$. Considering P_T as the overall transmission power emitted by BS, the design of $\mathbf{F}_b$ and $\mathbf{B}_b$ satisfies the power limitation, indicated by the equation $||\mathbf{F}_b \mathbf{B}_b||_F^2 = P_T$. Then, the received signal at the UAV relay is as follows:

$$\begin{aligned} \mathbf{y}_1 &= \mathbf{H}_1 \mathbf{s}_1 + \mathbf{n}_1, \\ &= \mathbf{H}_1 \mathbf{F}_b \mathbf{B}_b \mathbf{d} + \mathbf{n}_1, \end{aligned} \tag{4.3}$$

where $n_1 \in \mathbb{C}^{N_r}$ denotes the zero-mean complex symmetric Gaussian noise vector at UAV relay with covariance matrix $\mathbb{E}\{\mathbf{n}_1 \mathbf{n}_1^H\} = \sigma_n^2 \mathbf{I}_{N_r} \in \mathbb{C}^{N_r \times N_r}$. Then, after BB processing at the UAV relay, the received signal can be expressed as follows:

$$\tilde{\mathbf{y}}_1 = \mathbf{B}_{u,r} \mathbf{F}_{u,r} \mathbf{H}_1 \mathbf{F}_b \mathbf{B}_b \mathbf{d} + \mathbf{B}_{u,r} \mathbf{F}_{u,r} \mathbf{n}_1. \tag{4.4}$$

Consequently, as a DF relay, UAV uses $\tilde{\mathbf{y}}_1$ to first decode and then re-encode the information for transmission in the second time slot. Afterward, the signal transmitted by UAV is

$$\mathbf{s}_2 = \mathbf{F}_{u,t} \mathbf{B}_{u,t} \mathbf{P} \hat{\mathbf{d}}, \tag{4.5}$$

where $\hat{\mathbf{d}}$ is the re-encoded signal at the UAV relay. In the next step, the UAV transmits this signal using N_t antennas through channel $\mathbf{H}_2$ to users. The received signal by kth user can be expressed in the following way: The received signal by kth user in gth group can be expressed in the following way:

$$\begin{aligned} y_{g,k} &= \mathbf{h}_{2,g_k}^T \mathbf{s}_2 + n_{2,g_k} = \mathbf{h}_{2,g_k}^T \mathbf{F}_{u,t} \mathbf{B}_{u,t} \mathbf{P} \hat{\mathbf{d}} + n_{2,g_k} \\ &= \underbrace{\sqrt{p_{g_k}} \mathbf{h}_{2,g_k}^T \mathbf{F}_{u,t} \mathbf{b}_{u,t,g_k} \hat{d}_{g_k}}_{\text{Desired Signal}} + \underbrace{\sum_{\hat{k} \neq k}^{K_g} \sqrt{p_{g_{\hat{k}}}} \mathbf{h}_{2,g_k}^T \mathbf{F}_{u,t} \mathbf{b}_{u,t,g_{\hat{k}}} \hat{d}_{g_{\hat{k}}}}_{\text{Intra-group interference}} \\ &+ \underbrace{\sum_{q \neq g}^{G} \sum_{\hat{k}=1}^{K_g} \sqrt{p_{g_{\hat{q}}}} \mathbf{h}_{2,g_k}^T \mathbf{F}_{u,t} \mathbf{b}_{u,t,q_{\hat{k}}} \hat{d}_{q_{\hat{k}}}}_{\text{Inter-group interference}} + \underbrace{n_{g_k}}_{\text{Noise}}, \end{aligned} \tag{4.6}$$

where $g_k = k + \sum_{g'=1}^{g-1} K_{g'}$ represents the user index, $\mathbf{h}_{2,g_k} \in \mathbb{C}^{N_t}$ is the channel vector between UAV and respective user, and n_{g_k} indicates the complex circular symmetric Gaussian noise that follows the distribution $\mathcal{CN}(\mathbf{0}, \sigma^2)$. Subsequently, the AR of the first link, which is between BS and UAV, can be given as follows:

$$\mathrm{R}_1(\mathbf{F}_b, \mathbf{B}_b, \mathbf{F}_{u,r}, \mathbf{B}_{u,r}) = \log_2 |\mathbf{I}_K + \mathbf{Q}_1^{-1} \mathbf{B}_{u,r} \mathcal{H}_1 \mathbf{B}_b \mathbf{B}_b^H \mathcal{H}_1^H \mathbf{B}_{u,r}^H|, \tag{4.7}$$

where $\mathbf{Q}_1^{-1} = (\sigma_n^2 \mathbf{B}_{u,r} \mathbf{F}_{u,r})^{-1} \mathbf{F}_{u,r}^H \mathbf{B}_{u,r}^H$ and $\mathcal{H}_1 = \mathbf{F}_{u,r} \mathbf{H}_1 \mathbf{F}_b$. Similarly, the total AR for the second link, which is between UAV and users, is calculated based on the instantaneous SINR for the kth user in the gth group (SINR_{g_k}) as

$$\mathrm{SINR}_{g_k} = \frac{p_{g_k} |\mathbf{h}_{2,k}^H \mathbf{F}_{u,t} \mathbf{b}_{u,t,g_k}|^2}{\sum\limits_{\hat{k} \neq k}^{K_g} p_{g_{\hat{k}}} |\mathbf{h}_{2,k}^H \mathbf{F}_{u,t} \mathbf{b}_{u,t,g_{\hat{k}}}|^2 + \sum\limits_{q \neq g}^{G} \sum\limits_{\hat{k} \neq k}^{K_g} p_{q_{\hat{k}}} |\mathbf{h}_{2,k}^H \mathbf{F}_{u,t} \mathbf{b}_{u,t,q_{\hat{k}}}|^2 + \sigma_n^2}. \tag{4.8}$$

Utilizing the instantaneous SINR, the ergodic sum rate capacity of the second link R2, in UAV-assisted mmWave MU-mMIMO systems, is described as

$$\mathrm{R}_2(\mathbf{F}_{u,t}, \mathbf{B}_{u,t}, \mathbf{P}, x_u, y_u) = \mathbb{E} \left\{ \sum_{g=1}^{G} \sum_{k=1}^{K_g} \mathbb{E} \left[\log_2(1 + \mathrm{SINR}_{g_k}) \right] \right\}. \tag{4.9}$$

4.2.2 Channel Model

Although LoS channel models are useful for simple scenarios, they fall short of capturing the complexities of the channel, such as multipath fading and shadowing, especially for multiuser cases. Conversely, mmWave channel models offer a more precise representation of channel characteristics, accounting for the effects of NLoS paths and obstacles on signal propagation in UAV-assisted communications. Therefore, we utilize mmWave channels for both links. The channel between the BS and UAV is modeled using the Saleh-Valenzuela channel model [18], which is given as

$$\mathbf{H}_1 = \sum_{c=1}^{C}\sum_{l=1}^{L} z_{1_{cl}} \tau_{l_{cl}}^{-\eta} \boldsymbol{\phi}_1^{(r)}(\theta_{cl}^{(r)}, \psi_{cl}^{(r)}) \boldsymbol{\phi}_1^{(t)T}(\theta_{cl}^{(t)}, \psi_{cl}^{(t)}) = \boldsymbol{\Phi}_1^{(r)} \mathbf{Z}_1 \boldsymbol{\Phi}_1^{(t)}. \tag{4.10}$$

Here, C denotes the total number of clusters, L is the total number of paths from the BS to the UAV, $z_{1_{cl}}$ represents the complex path gain of the lth path in the cth cluster that follows $\mathcal{CN}(\mathbf{0}, \frac{1}{L})$, and η is the path loss exponent. Furthermore, $\boldsymbol{\phi}_1^{(k)}(.,.)$ denotes the corresponding transmit or receive array steering vector for a URA, which is specified as [19]

$$\begin{aligned}\boldsymbol{\phi}_1^{(k)}(\theta, \psi) =& [1, e^{-j2\pi d \sin(\theta)\cos(\psi)}, \cdots, e^{-j2\pi d(N_x-1)\sin(\theta)\cos(\psi)}] \\ &\otimes [1, e^{-j2\pi d \sin(\theta)\sin(\psi)}, \cdots, e^{-j2\pi d(N_y-1)\sin(\theta)\sin(\psi)}],\end{aligned} \tag{4.11}$$

where $k = \{r, t\}$, N_x, and N_y are the horizontal and vertical sizes of the antenna arrays at the BS and UAV, d is the spacing between elements, $\mathbf{Z}_1 = \text{diag}(z_{1,1}\tau_{1,1}^{-\eta}, \ldots, z_{1,L}\tau_{1,L}^{-\eta}) \in \mathbb{C}^{L\times L}$ denotes the diagonal gain matrix, and $\boldsymbol{\Phi}_1^{(r)} \in \mathbb{C}^{N_r\times L}$ and $\boldsymbol{\Phi}_1^{(t)} \in \mathbb{C}^{L\times N_t}$ are the phase response matrices for receiving and transmitting, respectively. Additionally, the angles $\theta_{cl}^{(t)} \in [\theta_c^{(t)} - \delta_c^{\theta(t)}, \theta_c^{(t)} + \delta_c^{\theta(t)}]$ and $\psi_{cl}^{(t)} \in [\psi_c^{(t)} - \delta_c^{\psi(t)}, \psi_c^{(t)} + \delta_c^{\psi(t)}]$ represent the EAoD and AAoD for the lth path in channel $\mathbf{H}_1$, respectively. Here, $\theta_c^{(t)}$ denotes the mean EAoD, with $\delta_c^{\theta(t)}$ indicating the EAoD spread, while $\psi_c^{(t)}$ is the mean AAoD, with $\delta_c^{\psi(t)}$ representing the AAoD spread. Similarly, the angles $\theta_{cl}^{(r)}$ within the range $[\theta_c^{(r)} - \delta_c^{\theta(r)}, \theta_c^{(r)} + \delta_c^{\theta(r)}]$ and $\psi_{cl}^{(r)}$ within $[\psi_c^{(r)} - \delta_c^{\psi(r)}, \psi_c^{(r)} + \delta_c^{\psi(r)}]$ correspond to the EAoA and AAoA, respectively. In this context, $\theta_c^{(r)}$ and $\psi_c^{(r)}$ are the mean EAoA and AAoA, while $\delta_c^{\theta(r)}$ and $\delta_c^{\psi(r)}$ denote the spreads of the elevation and azimuth angles, respectively. Then, the channel vector between the UAV and kth user can be expressed as

$$\mathbf{h}_{2,k}^T = \sum_{l=1}^{L} z_{2,k_l} \tau_{2,k_l}^{-\eta} \boldsymbol{\phi}(\theta_{k_l}, \psi_{k_l}) = \mathbf{z}_{2,k}^T \boldsymbol{\Phi}_{2,k} \in \mathbb{C}^{N_t}, \tag{4.12}$$

where $z_{2,k_l} \sim \mathcal{CN}(0, \frac{1}{L})$ is the complex path gain of lth path in the second link, and $\boldsymbol{\phi}(.,.) \in \mathbb{C}^{N_t}$ is the UAV downlink array phase response vector. As described in (4.12), the obtained downlink channel comprises two distinct parts: (1) a fast time-varying path gain vector $z_{2,k} = [z_{2,k_1}\tau_{2,k_1}^{-\eta}, \ldots, z_{2,k_L}\tau_{2,k_L}^{-\eta}]^T \in \mathbb{C}^L$; and (2) a slow time-varying downlink array phase response matrix $\boldsymbol{\Phi}_{2,k} \in \mathbb{C}^{L\times N_t}$ where each row is constituted by $\boldsymbol{\phi}(\theta_{kl}, \psi_{kl})$. Afterward, the channel matrix for the second link can be expressed as

$$\mathbf{H}_2 = [\mathbf{h}_{2,1}, \cdots, \mathbf{h}_{2,K}]^T = \mathbf{Z}_2\boldsymbol{\Phi}_2 \in \mathbb{C}^{K\times N_t}, \tag{4.13}$$

where $\mathbf{Z}_2 = [\mathbf{z}_{2,1}, \cdots, \mathbf{z}_{2,K}]^T \in \mathbb{C}^{K\times L}$ is the complete path gain matrix for all downlink users.

4.2.3 Problem Formulation

Considering that the UAV is deployed at a fixed altitude z_u, we define its 2-D deployment as $\mathbf{x}_u = [x_u, y_u]^T \in \mathbb{R}^2$ within a given flying span. The total transmission rate from the BS to users under DF protocol, $\mathrm{R_T} = (1/2)\min(\mathrm{R_1}, \mathrm{R_2})$, can be maximized through the joint optimization of the beamforming stages $\mathbf{B}_b, \mathbf{F}_b, \mathbf{B}_{u,r}, \mathbf{F}_{u,r}, \mathbf{B}_{u,t}$, and $\mathbf{F}_{u,t}$ along with the optimal PA matrix $\mathbf{P}$ and UAV positioning vector $\mathbf{x}_u$. Therefore, the optimization problem can be formulated as follows:

$$\begin{aligned}
&\max_{\{\mathbf{F}_b,\mathbf{B}_b,\mathbf{F}_{u,r},\mathbf{B}_{u,r},\mathbf{F}_{u,t},\mathbf{B}_{u,t},\mathbf{P},\mathbf{x}_u\}} \quad \mathrm{R_T} \\
&\text{s.t.} \quad C_1 : |\mathbf{F}_b(i,j)| = \frac{1}{\sqrt{N_T}} \quad \forall i, j \\
&\qquad C_2 : |\mathbf{F}_{u,r}(i,j)| = \frac{1}{\sqrt{N_r}}, |\mathbf{F}_{u,t}(i,j)| = \frac{1}{\sqrt{N_t}}, \quad \forall i, j \\
&\qquad C_3 : \mathbb{E}\left\{\|\mathbf{s}_1\|_2^2\right\} \le P_T \\
&\qquad C_4 : \mathbb{E}\left\{\|\mathbf{s}_2\|_2^2\right\} \le \sum_{k=1}^{K} p_k \mathbf{b}_{u,t,k}^H \mathbf{F}_{u,t}^H \mathbf{F}_{u,t} \mathbf{F}_{u,t} \mathbf{b}_{u,t,k}, \le P_T \\
&\qquad C_5 : p_k \ge 0 \quad \forall k \\
&\qquad C_6 : x_{\min} \le x_u \le x_{\max}, y_{\min} \le y_u \le y_{\max},
\end{aligned} \tag{4.14}$$

where C_1 and C_2 are related to the CM constraint for the BS and UAV, respectively, C_3 and C_4 indicate the transmit power constraint at the BS and UAV, respectively, C_5 represents the nonnegative allocated power to each node, and C_6 implies the

UAV positioning within the given flying span. Here, $[x_{\min}, x_{\max}]$ and $[y_{\min}, y_{\max}]$ are the UAV relay deployment range in the x-axis and y-axis, respectively. The optimization problem defined in (4.14) is non-convex and intractable due to the following reasons: (1) the CM constraint at each RF stage and (2) fractional programming variables are entangled with each other. To solve this challenging problem, we propose three different PSO-based algorithmic solutions in Sect. 4.3, which can achieve a near-optimal solution in finding optimal PA $\mathbf{P}$ and UAV deployment $\mathbf{x}_u$.

4.3 Joint HBF, PA, and UAV Location Optimization

In this section, our objectives are to reduce the CSI overhead and the number of RF chains, while mitigating the inter-user interference to maximize the total AR of a dual-hop UAV-assisted MU-mMIMO communications system. In this regard, we consider the joint optimization of the UAV location, PA to multiple users, and HBF design for BS and UAV. Since the optimization problem is intractable, we first design $\mathbf{F}_b, \mathbf{B}_b, \mathbf{F}_{u,t}, \mathbf{F}_{u,r}, \mathbf{B}_{u,t}, \mathbf{B}_{u,r}$ based on some fixed UAV location and then reformulate RF and BB stages for the optimal UAV location as well as adjust the allocated power in the MU PA block $\mathbf{P}$ by using three different algorithmic schemes: (1) optimal PA for fixed UAV location; (2) UAV location optimization for equal PA; and (3) joint UAV location optimization and optimal PA.

4.3.1 RF Stage Design

The design of the RF stages $\mathbf{F}_b$, $\mathbf{F}_{u,r}$, and $\mathbf{F}_{u,t}$ aims to optimize the beamforming gain for intended signals, as indicated in (4.4) and (4.6). Utilizing (4.10), the efficient channel for the first link is formulated as follows:

$$\mathcal{H}_1 = \mathbf{F}_{u,r}\mathbf{H}_1\mathbf{F}_b = \mathbf{F}_{u,r}\boldsymbol{\Phi}_1^{(r)}\mathbf{Z}_1\boldsymbol{\Phi}_1^{(t)}\mathbf{F}_b. \tag{4.15}$$

In order to maximize the gain for transmit (receive) beamforming and fully harness the DoF provided by the first channel, the columns of $\mathbf{F}_b(\mathbf{F}_{u,r})$ should belong to the subspace spanned by $\boldsymbol{\Phi}_1^{(t)}(\boldsymbol{\Phi}_1^{(r)})$. Therefore, the conditions $\text{Span}(\mathbf{F}_b) \subset \text{Span}(\boldsymbol{\Phi}_1^{(t)})$ and $\text{Span}(\mathbf{F}_{u,r}) \subset \text{Span}(\boldsymbol{\Phi}_1^{(r)})$ should be met. It is important to note here that the transmit (receive) phase response matrix $\boldsymbol{\Phi}_1^{(t)}(\boldsymbol{\Phi}_1^{(r)})$ is determined by the slow time-varying AoA and AoD information. Consequently, the AoD and AoA supports for the first channel are defined as follows:

$$\text{AoD} = \left\{\sin(\theta)[\cos(\psi), \sin(\psi)] | \theta \in \boldsymbol{\theta}_1^{(t)}, \psi \in \boldsymbol{\psi}_1^{(t)}\right\}, \tag{4.16}$$

$$\text{AoA} = \left\{\sin(\theta)[\cos(\psi), \sin(\psi)] | \theta \in \boldsymbol{\theta}_1^{(r)}, \psi \in \boldsymbol{\psi}_1^{(r)}\right\}, \tag{4.17}$$

where $\boldsymbol{\theta}_1^{(t)} = \cup_{c=1}^{C}[\theta_{1,c}^{(t)} - \delta_{1,c}^{(t)\theta}, \theta_{1,c}^{(t)} + \delta_{1,c}^{(t)\theta}]$ and $\boldsymbol{\psi}_1^{(t)} = \cup_{c=1}^{C}[\psi_{1,c}^{(t)} - \delta_{1,c}^{(t)\psi}, \psi_{1,c}^{(t)} + \delta_{1,c}^{(t)\psi}]$ are the elevation and azimuth AoD supports of the first channel, respectively. In the same fashion, $\boldsymbol{\theta}_1^{(r)} = \cup_{c=1}^{C}[\theta_{1,c}^{(r)} - \delta_{1,c}^{(r)\theta}, \theta_{1,c}^{(r)} + \delta_{1,c}^{(r)\theta}]$ and $\boldsymbol{\psi}_1^{(r)} = \cup_{c=1}^{C}[\psi_{1,c}^{(r)} - \delta_{1,c}^{(r)\psi}, \psi_{1,c}^{(r)} + \delta_{1,c}^{(r)\psi}]$ denote the elevation and azimuth AoA supports of the channel between BS and UAV, respectively. Subsequently, the transmit RF beamformer $\mathbf{F}_b$ is assembled using the transmit steering vector in the following manner:

$$\mathbf{e}_b^{(t)}(\theta, \psi) = \frac{1}{\sqrt{N_{x,T}}}[1, e^{j2\pi d \sin(\theta)\cos(\psi)}, \cdots, e^{j2\pi d(N_{x,T}-1)\sin(\theta)\cos(\psi)}]^T \otimes \frac{1}{\sqrt{N_{y,T}}}[1, e^{j2\pi d \sin(\theta)\sin(\psi)}, \cdots, e^{j2\pi d(N_{y,T}-1)\sin(\theta)\sin(\psi)}]^T. \tag{4.18}$$

In order to minimize the use of RF chains while encompassing the entire AoD and AoA supports, quantized angle pairs are defined as: $\lambda_{x,b}^{n(t)} = -1 + \frac{2n-1}{N_{x,T}}$ for $n = 1, \cdots, N_{x,T}$ and $\lambda_{y,b}^{k(t)} = -1 + \frac{2k-1}{N_{y,T}}$ for $k = 1, \cdots, N_{y,T}$, where $N_{x,T}$ and $N_{y,T}$ are the antenna elements along x- and y-axis, respectively. Subsequently, the quantized angle pairs inside the AoD support meeting the criteria in (4.16) are derived as

$$\left(\lambda_{x,b}^{n(t)}, \lambda_{y,b}^{k(t)}\right) | \sin(\theta)\cos(\psi) \in \boldsymbol{\lambda}_{x,b}^{n(t)}, \sin(\theta)\sin(\psi) \in \boldsymbol{\lambda}_{y,b}^{k(t)}, \tag{4.19}$$

where $\boldsymbol{\lambda}_{x,b}^{n(t)} = [\lambda_{x,b}^{n(t)} - \frac{1}{N_{x,T}}, \lambda_{x,b}^{n(t)} + \frac{1}{N_{x,T}}]$ and $\boldsymbol{\lambda}_{y,b}^{k(t)} = [\lambda_{y,b}^{k(t)} - \frac{1}{N_{y,T}}, \lambda_{y,b}^{k(t)} + \frac{1}{N_{y,T}}]$ are the boundary of $\lambda_{x,b}^{n(t)}$ and $\lambda_{y,b}^{k(t)}$, respectively. Then, $\mathbf{F}_b$ can be expressed in the form of transmit steering vector using (4.16) and (4.19) as

$$\mathbf{F}_b = \left[\mathbf{e}_b^{(t)}(\lambda_{x,b}^{n_1(t)}, \lambda_{y,b}^{k_1(t)}), \cdots, \mathbf{e}_b^{(t)}(\lambda_{x,b}^{n_{N_{RF_b}}(t)}, \lambda_{y,b}^{k_{N_{RF_b}}(t)})\right] \in \mathbb{C}^{N_T \times N_{RF_b}}. \tag{4.20}$$

In a similar manner, the design of the UAV's receive RF stage can be implemented using the receive steering vector in the following way:

$$\mathbf{e}_u^{(r)}(\theta, \psi) = \frac{1}{\sqrt{N_{x,r}}}[1, e^{-j2\pi d \sin(\theta)\cos(\psi)}, \cdots, e^{-j2\pi d(N_{x,r}-1)\sin(\theta)\cos(\psi)}]^T \otimes \frac{1}{\sqrt{N_{y,r}}}[1, e^{-j2\pi d \sin(\theta)\sin(\psi)}, \cdots, e^{-j2\pi d(N_{y,r}-1)\sin(\theta)\sin(\psi)}]^T. \tag{4.21}$$

Then, the UAV receive RF stage can be expressed using quantized angle pairs $\lambda_{x,u}^{n(r)} = -1+\frac{2n-1}{N_{x,r}}$ for $n = 1, \cdots, N_{x,r}$ and $\lambda_{y,u}^{k(r)} = -1+\frac{2k-1}{N_{y,r}}$ for $k = 1, \cdots, N_{y,r}$ as

$$\mathbf{F}_{u,r}=\left[\mathbf{e}_u^{(r)}(\lambda_{x,u}^{n_1(r)}, \lambda_{y,u}^{k_1(r)}), \cdots, \mathbf{e}_u^{(r)}(\lambda_{x,u}^{n_{N_{RF_u}}(r)}, \lambda_{y,u}^{k_{N_{RF_u}}(r)})\right]^T \in \mathbb{C}^{N_{RF_u}\times N_r}. \quad (4.22)$$

Here, the quantized angle pairs inside the AoA support meeting the criteria in (4.17) can be written as

$$\left(\lambda_{x,u}^{n(r)}, \lambda_{y,u}^{k(r)}\right) |\sin(\theta)\cos(\psi) \in \boldsymbol{\lambda}_{x,u}^{n(r)}, \sin(\theta)\sin(\psi) \in \boldsymbol{\lambda}_{y,u}^{k(r)}, \quad (4.23)$$

where $\boldsymbol{\lambda}_{x,u}^{n(r)} = [\lambda_{x,u}^{n(r)} - \frac{1}{N_{x,r}}, \lambda_{x,u}^{n(r)} + \frac{1}{N_{x,r}}]$ and $\boldsymbol{\lambda}_{y,u}^{k(r)} = [\lambda_{y,u}^{k(r)} - \frac{1}{N_{y,r}}, \lambda_{y,u}^{k(r)} + \frac{1}{N_{y,r}}]$ are the boundary of $\lambda_{x,u}^{n(r)}$ and $\lambda_{y,u}^{k(r)}$, respectively. After the design of $\mathbf{F}_b$ and $\mathbf{F}_{u,r}$ RF stages, the UAV transmit is designed to support K users that are clustered into G groups based on their AoD information. Here, each group g contains K_g number of users such that $K = \sum_{g=1}^{G} K_g$. The index $g_k = \sum_{g'=1}^{g-1} +k$ is used to denote the kth user in the group g. According to the user groups, we design G different subblocks for the UAV transmit RF stage as $\mathbf{F}_{u,t} == \left[\mathbf{F}_{u,t,1}, \cdots, \mathbf{F}_{u,t,G}\right] \in \mathbb{C}^{N_t\times N_{RF_u}}$, where $\mathbf{F}_{u,t,g} \in \mathbb{C}^{N_t\times N_{RF_{u,g}}}$ is the RF beamformer for group g in such that $N_{RF_u} = \sum_{g=1}^{G} N_{RF_{u,g}}$. Then, we define $\mathcal{H}_2 = \mathbf{H}_2\mathbf{F}_{u,t} \in \mathbb{C}^{K\times N_{RF_u}}$ as the reduced-size effective channel matrix observed from UAV transmit BB stage $\mathbf{B}_{u,t}$. Using $\mathbf{H}_2 = [\mathbf{H}_{2,1}^T, \cdots, \mathbf{H}_{2,G}^T]^T \in \mathbb{C}^{K\times N_t}$, we can write the effective channel matrix for the second link as follows:

$$\mathcal{H}_2 = \begin{bmatrix} \mathbf{H}_{2,1}\mathbf{F}_{u,t,1} & \mathbf{H}_{2,1}\mathbf{F}_{u,t,2} & \cdots & \mathbf{H}_{2,1}\mathbf{F}_{u,t,G} \\ \mathbf{H}_{2,2}\mathbf{F}_{u,t,1} & \mathbf{H}_{2,2}\mathbf{F}_{u,t,2} & \cdots & \mathbf{H}_{2,2}\mathbf{F}_{u,t,G} \\ \vdots & \vdots & \ddots & \vdots \\ \mathbf{H}_{2,G}\mathbf{F}_{u,t,1} & \mathbf{H}_{2,G}\mathbf{F}_{u,t,2} & \cdots & \mathbf{H}_{2,G}\mathbf{F}_{u,t,G} \end{bmatrix}, \quad (4.24)$$

where the diagonal block matrix $\mathcal{H}_{2,g} = \mathbf{H}_{2,g}\mathbf{F}_{u,t,g} = \mathbf{Z}_{2,g}\mathbf{\Phi}_{2,g}\mathbf{F}_{u,t,g}$ is the effective channel matrix for group g and off-diagonal block matrix $\mathcal{H}_{2,\hat{g}} = \mathbf{H}_{2,g}\mathbf{F}_{u,t,g} = \mathbf{Z}_{2,g}\mathbf{\Phi}_{2,g}\mathbf{F}_{u,t,g}$ is the effective interference channel matrix between groups g and $\hat{g}$, $\forall \hat{g} \neq g$. The RF beamformer matrices are designed to eliminate intergroup interference as follows:

$$\mathbf{\Phi}_{2,\hat{g}}\mathbf{F}_{u,t,g} \approx 0, \;\; \forall \hat{g} \neq g \text{ and } \hat{g}, g = 1, \cdots, G. \quad (4.25)$$

To design $\mathbf{F}_{u,t}$, which can satisfy the above zero condition, the columns of $\mathbf{F}_{u,t,g}$ should belong to the intersection of the null spaces of $\mathbf{\Phi}_{2,\hat{g}}$, i.e., $\text{Span}\left(\mathbf{F}_{u,t,g}\right) \subset \cap_{\hat{g}\neq g}\text{Null}\left(\mathbf{\Phi}_{2,\hat{g}}\right)$. Moreover, in order to maximize the beamforming gain, the columns of $\mathbf{F}_{u,t,g}$ should belong to the subspace spanned by $\boldsymbol{\Phi}_{2,g}$, i.e., $\text{Span}\left(\mathbf{F}_{u,t,g}\right) \subset \text{Null}\left(\boldsymbol{\Phi}_{2,g}\right)$. Thus, the intersection of $\text{Span}(\boldsymbol{\Phi}_{2,g})$ and $\text{Null}(\boldsymbol{\Phi}_{2,\hat{g}})$ $\forall \hat{g} \neq g$ should not be empty to obtain the RF beamformer matrix satisfying the above conditions. Similar to the design of $\mathbf{F}_b$, the AoD support of the group g can be expressed as the union of AoD supports for all users in the corresponding group as follows:

$$\text{AoD}_g = \left\{\sin(\theta)[\cos(\psi), \sin(\psi)] | \theta \in \boldsymbol{\theta}_{2,g}, \psi \in \boldsymbol{\psi}_{2,g}\right\}, \tag{4.26}$$

where $\boldsymbol{\theta}_{2,g} = \left[\theta_{2,g}^{\min}, \theta_{2,g}^{\max}\right] = \left[\min_{2,g_k,\hat{g}} \theta_{2,g_k,\hat{g}}, \max_{2,g_k,\hat{g}} \theta_{2,g_k,\hat{g}}\right]$ is the elevation angle support for group g and $\boldsymbol{\psi}_{2.g} = \left[\psi_{2,g}^{\min}, \psi_{2,g}^{\max}\right] = \left[\min_{2,g_k,\hat{g}} \psi_{2,g_k,\hat{g}}, \max_{2,g_k,\hat{g}} \psi_{2,g_k,\hat{g}}\right]$ is the azimuth angle support for group g. To achieve $\text{Span}\left(\mathbf{F}_{u,t,g}\right) \subset \text{Span}\left(\boldsymbol{\Phi}_{2,g}\right)$, the columns of the UAV transmit RF beamformer for the group g can be constructed as

$$\mathbf{F}_{u,t,g} = \{\mathbf{e}_u^{(t)}(\varphi, \vartheta) \,|\, (\varphi, \vartheta) \in \text{AoD}_g\}, \tag{4.27}$$

where $\varphi = \sin\theta \cos\psi$ and $\vartheta = \sin\theta \sin\psi$ represent the beam directions for the UAV transmit steering vector for URA, which is defined as follows:

$$\mathbf{e}_u^{(t)}(\varphi, \vartheta) = \mathbf{e}_{x,u}^{(t)}(\varphi) \otimes \mathbf{e}_{y,u}^{(t)}(\vartheta). \tag{4.28}$$

Then, we can construct the UAV transmit RF beamformer for each group g based on the quantized angle pairs as

$$\mathbf{F}_{u,t,g} = \left[\mathbf{e}_u^{(t)}(\lambda_{x,u}^{n_1(t)}, \lambda_{y,u}^{k_1(t)}), \cdots, \mathbf{e}_u^{(t)}(\lambda_{x,u}^{n_{N_{RF_{u,g}}}(t)}, \lambda_{y,u}^{k_{N_{RF_{u,g}}}(t)})\right]^T \in \mathbb{C}^{N_t \times N_{RF_{u,g}}}, \tag{4.29}$$

where $\boldsymbol{\lambda}_{x,u}^{n_i(t)} = [\lambda_{x,u}^{n_i(t)} - \frac{1}{N_{x,t}}, \lambda_{x,u}^{n_i(t)} + \frac{1}{N_{x,t}}]$ and $\boldsymbol{\lambda}_{y,u}^{k_i(t)} = [\lambda_{y,u}^{k_i(t)} - \frac{1}{N_{y,t}}, \lambda_{y,u}^{k_i(t)} + \frac{1}{N_{y,t}}]$ are the boundary of $\lambda_{x,u}^{n_i(t)}$ and $\lambda_{y,u}^{k_i(t)}$, respectively. The BS and UAV RF stage designs do not require the fast time-varying instantaneous CSI and are only based on the slow time-varying AoD/AoA information. Particularly, the design of BS and UAV RF beamformers requires only four angular parameters, which are the mean of elevation and azimuth AoD (AoA) and their angular spread.

4.3.2 BB Stage Design

The next step is to design the BB stages $\mathbf{B}_b$ and $\mathbf{B}_{u,r}$ according to the effective channel matrix $\mathcal{H}_1$ as it is given in (4.15). Utilizing SVD, we can write

$$\mathcal{H}_1 = \mathcal{Q}_1 \boldsymbol{\Sigma}_1 \mathcal{V}_1^H. \tag{4.30}$$

Here, $\mathcal{Q}_1 \in \mathbb{C}^{N_{RF_u} \times rank(\mathcal{H}_1)}$ and $\mathcal{V}_1 \in \mathbb{C}^{N_{RF_b} \times rank(\mathcal{H}_1)}$ are tall unitary matrices, and $\mathbf{\Sigma}_1 = \text{diag}(\sigma_1^2, \cdots, \sigma_{rank(\mathcal{H}_1)}^2) \in \mathbb{C}^{rank(\mathcal{H}_1) \times rank(\mathcal{H}_1)}$ is a diagonal matrix containing the singular values arranged in descending order. Considering rank$(\mathcal{H}_1) \geq K$, $\mathcal{V}_1$ can be divided into two parts: $\mathcal{V}_1 = [\mathcal{V}_{1,1}, \mathcal{V}_{1,2}]$, where $\mathcal{V}_{1,1} \in \mathbb{C}^{N_{RF_b} \times K}$. Subsequently, the BB stages $\mathbf{B}_b$ and $\mathbf{B}_{u,r}$ for the BS and UAV can be written as

$$\mathbf{B}_b = \sqrt{\frac{P_T}{K}} \mathcal{V}_1 \in \mathbb{C}^{N_{RF_b} \times K}, \tag{4.31}$$

$$\mathbf{B}_{u,r} = \mathcal{Q}_1^H \in \mathbb{C}^{K \times N_{RF_b}}. \tag{4.32}$$

Similarly, the reduced-size effective CSI $\mathcal{H}_2$ given in (4.24) is employed for the UAV transmit BB stage design. We consider joint-group-processing (JGP) technique as designed in [19–21]. The design of BB stage $\mathbf{B}_{u,t}$ not only reduces the intragroup interference but also mitigates the residual intergroup interference remaining after RF beamformer design. By applying the RZF technique, $\mathbf{B}_{u,t}$ is defined as follows:

$$\mathbf{B}_{u,t} = \left(\mathcal{H}_2^H \mathcal{H}_2 + \xi N_{RF_u} \mathbf{I}_{N_{RF_u}}\right)^{-1} \mathcal{H}_2^H \in \mathbb{C}^{N_{RF_u} \times K}, \tag{4.33}$$

where $\xi = \frac{\sigma^2}{P_T}$ is the regularization parameter and $\mathbf{I}_{N_{RF_u}} \in \mathbb{C}^{N_{RF_u} \times N_{RF_u}}$.

4.3.3 UAV Deployment and MU PA

After the formulation of RF beamformers $\mathbf{F}_b$, $\mathbf{F}_{u,r}$, $\mathbf{F}_{u,t}$, and BB stages $\mathbf{B}_b$, $\mathbf{B}_{u,t}$, $\mathbf{B}_{u,r}$, the optimization problem for maximum AR can be formulated as follows:

$$\begin{aligned}
\max_{\{\mathbf{P}, \mathbf{x}_o\}} \quad & \mathrm{R}_T(\mathbf{F}_b, \mathbf{B}_b, \mathbf{F}_{u,t}, \mathbf{B}_{u,t}, \mathbf{F}_{u,r}, \mathbf{B}_{u,r}, \mathbf{P}, \mathbf{x}_u) \\
\text{s.t.} \quad & C_4: \ \mathbb{E}\{\|\mathbf{s}_2\|_2^2\} = \sum\nolimits_{k=1}^{K} p_k \mathbf{b}_{u,t,k}^H \mathbf{F}_{u,t}^H \mathbf{F}_{u,t} \mathbf{b}_{u,t,k} \leq P_T, \\
& C_5: \ p_k \geq 0, \ \forall k, \\
& C_6: \ \mathbf{x}_{\min} \leq \mathbf{x}_u \leq \mathbf{x}_{\max},
\end{aligned} \tag{4.34}$$

where R_T is defined in (4.14). Even though the CM constraints for the RF beamformers and transmit power constraint for BS (i.e., C_1, C_2 and C_3 given in (4.14)) are satisfied via the RF and BB stages developed in Sects. 4.3.1 and 4.3.2, the updated optimization problem in (4.34) is still non-convex due to the joint dependence of both the allocated powers p_k, and the UAV location $\mathbf{x}_u = [x_u, y_u]^T$ on the SINR expression in (4.8), which is used in the sum rate R_2 calculation as given in (4.9). To overcome this challenge, we propose different PSO-based algorithmic solutions, which employ multiple agents, called particles, to explore the search space of the objective function given in (4.34). Initially, Z_p particles are randomly placed in the search space, where each particle communicates with other particles to share their personal best solution and update the current global best solution for the objective function. The particles then move iteratively for T iterations to reach the global optimum solution. Our PSO-based algorithms rely on two components, deterministic and stochastic, to guide the motion of the particles. The deterministic component utilizes knowledge from global and personal best solutions, while the stochastic component involves random movements. The proposed PSO-based algorithmic schemes are as follows: (1) optimal PA for fixed UAV location; (2) UAV positioning for equal PA; and (3) joint UAV positioning and optimal PA.

4.3.3.1 Optimal PA for Fixed UAV Location

In this problem, we allocate optimal MU power values $\mathbf{P}$ for a fixed UAV location $\mathbf{x}_u$. Then, the ith particle at the tth iteration represents an instance of the PA matrix as follows:

$$\mathbf{P}_i^{(t)} = \text{diag}(\sqrt{p_{1,i}^{(t)}}, \cdots, \sqrt{p_{K,i}^{(t)}}) \in \mathbb{R}^{K \times K}, \tag{4.35}$$

where $i = 1, \cdots, Z_p$ and $t = 0, 1, \cdots, T$ are the particle index and iteration index, respectively. The objective function $\text{R}_T(\mathbf{F}_b, \mathbf{B}_b, \mathbf{F}_{u,t}, \mathbf{B}_{u,t}, \mathbf{F}_{u,r}, \mathbf{B}_{u,r}, \mathbf{P}_i^{(t)}, \mathbf{x}_u)$ is calculated after evaluating the personal best and communicating for the global best solutions for each ith particle. The particles move from $\mathbf{P}_i^{(t)}$ to $\mathbf{P}_i^{(t+1)}$ until the maximum number of iterations is reached. To satisfy the transmit power constraints C_4 and C_5 given in (4.34), we define the following:

$$\kappa_i^{(t)} = \sqrt{\frac{P_T}{\mathbb{E}\left\{ \left\| \mathbf{F}_{u,t} \mathbf{B}_{u,t} \hat{\mathbf{P}}_i^{(t)} \hat{\mathbf{d}} \right\|_2^2 \right\}}} \overset{(a)}{=} \sqrt{\frac{P_T}{\sum_{k=1}^{K} \hat{p}_{k,i}^{(t)} \mathbf{b}_{u,t,k}^H \mathbf{b}_{u,t,k}}}, \tag{4.36}$$

where (a) follows the unitary property of UAV transmit beamformer (i.e., $\mathbf{F}_{u,t}^H \mathbf{F}_{u,t} = \mathbf{I}_{N_{RF_u}}$). The normalized PA matrix is given as: $\hat{\mathbf{P}}_i^{(t)} = \text{diag}(\sqrt{\hat{p}_{1,i}^{(t)}}, \cdots, \sqrt{\hat{p}_{K,i}^{(t)}}) \in \mathbb{R}^{K \times K}$ with $\hat{p}_{k,i}^{(t)} \in [0, 1]$. Here, $\mathbf{P}_i^{(t)}$ satisfies the transmit power constraints for any $\hat{\mathbf{P}}_i^{(t)}$ by defining $\mathbf{P}_i^{(t)} = \kappa_i^{(t)} \hat{\mathbf{P}}_i^{(t)}$. The PSO-based optimal PA solution is defined by two variables: the position $\hat{\mathbf{P}}_i^{(t)} \in \mathbb{R}^{K \times K}$ and velocity $\mathbf{W}_i^{(t)} \in \mathbb{R}^{K \times K}$. Initially, the diagonal entries of $\hat{\mathbf{P}}_i^{(0)}$ are uniformly distributed over the range [0, 1], while the velocity is set as $\mathbf{W}_i^{(0)} = 0$. During the iterations, the velocity and position matrices of ith particle are updated as follows:

$$\mathbf{W}_i^{(t+1)} = \gamma_1 \mathbf{Y}_1^{(t)} (\hat{\mathbf{P}}_{\text{best}}^{(t)} - \hat{\mathbf{P}}_i^{(t)}) + \gamma_2 \mathbf{Y}_2^{(t)} (\hat{\mathbf{P}}_{\text{best},i}^{(t)} - \hat{\mathbf{P}}_i^{(t)}) + \gamma_3^{(t)} \mathbf{W}_i^{(t)}, \tag{4.37}$$

$$\hat{\mathbf{P}}_i^{(t+1)} = \text{clip}(\hat{\mathbf{P}}_i^{(t)} + \text{clip}(\mathbf{W}_i^{(t+1)}, [w_{\min}, w_{\max}]), [0, 1]), \tag{4.38}$$

where γ_1 and γ_2 are the learning parameters for the global best $\mathbf{P}_{best}^{(t)}$ and the personal best $\mathbf{P}_{\text{best},i}^{(t)}$, $\gamma_3^{(t)} = \Gamma_l - \frac{t}{T}(\Gamma_u - \Gamma_l)$ is the inertia parameters with the upper bound Γ_u and lower bound Γ_l for decreasing the velocity as the number of iterations increases, $\mathbf{Y}_1^{(t)}$ and $\mathbf{Y}_2^{(t)}$ are the random diagonal matrices with the uniformly distributed entries over [0, 1], and $w_{\min}$ and $w_{\max}$ denote the minimum and maximum acceptable velocity for the particles, respectively. Here, clip $(x, [a, b]) = \max(a, \min(x, b))$ is used to prevent exceeding the maximum/minimum acceptable velocity and normalized power. Then, at tth iteration, the personal best of the ith particle and the global best over Z_p particles are respectively obtained as follows:

$$\hat{\mathbf{P}}_{\text{best},i}^{(t)} = \underset{\hat{\mathbf{P}}_i^{(t^*)}, \forall t^*=0,1,\cdots,t}{\arg\max} \; R_T(\mathbf{F}_b, \mathbf{B}_b, \mathbf{F}_{u,t}, \mathbf{B}_{u,t}, \mathbf{F}_{u,r}, \mathbf{B}_{u,r}, \kappa_i^{(t^*)} \hat{\mathbf{P}}_i^{(t^*)}, \mathbf{x}_u), \tag{4.39}$$

$$\hat{\mathbf{P}}_{best}^{(t)} = \underset{\hat{\mathbf{P}}_{\text{best},i}^{(t)}, \forall i=0,1,\cdots,Z_p}{\arg\max} \; R_T(\mathbf{F}_b, \mathbf{B}_b, \mathbf{F}_{u,t}, \mathbf{B}_{u,t}, \mathbf{F}_{u,r}, \mathbf{B}_{u,r}, \kappa_{\text{best},i}^{(t)} \hat{\mathbf{P}}_{\text{best},i}^{(t)}, \mathbf{x}_u). \tag{4.40}$$

Finally, the MU PA matrix can be derived as follows:

$$\mathbf{P} = \kappa_{\text{best}}^{(T)} \hat{\mathbf{P}}_{\text{best}}^{(T)}, \tag{4.41}$$

where $\kappa_{\text{best}}^{(t)}$ can be calculated by substituting $\hat{\mathbf{P}}_{\text{best}}^{(T)}$ into (4.36). The summary of the proposed PSO-based optimal PA for fixed UAV location with HBF is outlined in Algorithm 4.1.

4.3.3.2 UAV Positioning for Equal PA

In this problem, we optimize UAV location $\mathbf{x}_u = [x_u, y_u]^T$ in a given deployment span for fixed MU PA matrix $\mathbf{P}$, which is defined as

$$\mathbf{P} = \varepsilon \mathbf{I}_K \in \mathbb{C}^{K \times K}, \tag{4.42}$$

where ε is the normalization factor used to satisfy the power constraints C_4, C_5, and can be written as follows:

$$\varepsilon = \sqrt{\frac{P_T}{\text{tr}\mathbb{E}\{\hat{\mathbf{d}}\mathcal{H}_2^H \mathbf{T}^H \mathbf{F}_{u,t}^H \mathbf{F}_{u,t} \mathbf{T} \mathcal{H}_2\}}}, \tag{4.43}$$

Algorithm 4.1: Proposed joint HBF and optimal PA for fixed UAV location (J-HBF-PSOPA-FL) algorithm

Input: Z_p, T, (θ, ϕ), (x_b, y_b, z_b), (x_u, y_u, z_u).
Output: $\mathbf{P}$, $\mathbf{F}_b$, $\mathbf{B}_b$, $\mathbf{F}_{u,t}$, $\mathbf{B}_{u,t}$, $\mathbf{F}_{u,r}$, $\mathbf{B}_{u,r}$.
Formulate BS RF and BB stages using (4.20), (4.31).
Formulate UAV receive HBF stages using (4.32), (4.22).
Construct UAV transmit HBF stages via (4.33), (4.29).
for $i = 1 : Z_p$ **do**
  Initialize the velocity as $\mathbf{W}_i^{(0)} = \mathbf{0}$.
  Each entry of $\hat{\mathbf{P}}_i^{(0)}$ is uniformly distributed in [0, 1].
  Set the personal best $\hat{\mathbf{P}}_{\text{best},i}^{(0)} = \hat{\mathbf{P}}_i^{(0)}$.
end
Find the global best $\hat{\mathbf{P}}_{\text{best}}^{(0)}$ as in (4.40).
for $t = 1 : T$ **do**
  for $i = 1 : Z_p$ **do**
    Update the velocity $\mathbf{W}_i^{(t)}$ as in (4.37).
    Update the position $\hat{\mathbf{P}}_i^{(t)}$ as in (4.38).
    Find the personal best $\hat{\mathbf{P}}_{\text{best},i}^{(t)}$ as in (4.39).
  end
  Find the global best $\hat{\mathbf{P}}_{\text{best}}^{(t)}$ as in (4.40).
end
$\mathbf{P} = \kappa_{\text{best}}^{(T)} \hat{\mathbf{P}}_{\text{best}}^{(T)}$

where $\mathbf{T} = (\mathcal{H}_2^H \mathcal{H}_2 + \xi N_{RF_u} \mathbf{I}_{N_{RF_u}})^{-1}$. Since the number of feasible UAV positions required to search space is $\left(\frac{x_{\max} - x_{\min}}{\Delta_x}\right) \times \left(\frac{y_{\max} - y_{\min}}{\Delta_y}\right)$, where Δ_x and Δ_y represent the search space resolutions, we propose a PSO-based algorithmic solution to find optimal UAV position, while maximizing R_T. Following a similar approach as in Sect. 4.3.3, the ith particle at the tth iteration now represents an instance of the UAV location as

$$\mathbf{X}_i^{(t)} = [x_i^{(t)}, y_i^{(t)}]^T \in \mathbb{R}^2. \tag{4.44}$$

Here, the corresponding particle i represents the candidate UAV position and calculates the objective function as $R_T(\mathbf{F}_b, \mathbf{B}_b, \mathbf{F}_{u,t}, \mathbf{B}_{u,t}, \mathbf{F}_{u,r}, \mathbf{B}_{u,r}, \mathbf{P}, \mathbf{X}_i^{(t)})$. Then, the position $\mathbf{X}_i^{(t)}$ and velocity $\mathbf{V}_i^{(t)}$ are updated as follows:

$$\mathbf{X}_i^{(t+1)} = \mathbf{X}_i^{(t)} + \mathbf{V}_i^{(t+1)}, \tag{4.45}$$

$$\mathbf{V}_i^{(t+1)} = \gamma_1 \mathbf{Y}_1^{(t)} (\hat{\mathbf{X}}_{\text{best}}^{(t)} - \hat{\mathbf{X}}_i^{(t)}) + \gamma_2 \mathbf{Y}_2^{(t)} (\hat{\mathbf{X}}_{\text{best},i}^{(t)} - \hat{\mathbf{X}}_i^{(t)}) + \gamma_3^{(t)} \mathbf{V}_i^{(t)}. \tag{4.46}$$

Algorithm 4.2: Proposed joint HBF and UAV deployment for equal PA (J-HBF-PSOL-EQPA) algorithm

Input: $Z_p, T, (\theta, \psi), (x_1, y_1, z_1), (x_u, y_u, z_u)$.
Output: $\mathbf{x}_o, \mathbf{P}, \mathbf{F}_b, \mathbf{B}_b, \mathbf{F}_{u,t}, \mathbf{B}_{u,t}, \mathbf{F}_{u,r}, \mathbf{B}_{u,r}$.
Formulate BS RF and BB stages using (4.20), (4.31).
Formulate UAV receive HBF stages using (4.22), (4.32).
Construct UAV transmit HBF stages via (4.29), (4.33).
for $i = 1 : Z_p$ **do**
    Initialize the velocity as $\mathbf{V}_i^{(0)} = \mathbf{0}$.
    Each entry of $\mathbf{X}_i^{(0)}$ is uniformly distributed in [0, 1].
    Set the personal best $\mathbf{X}_{\text{best},i}^{(0)} = \mathbf{X}_i^{(0)}$.
end
Find the global best $\mathbf{X}_{\text{best}}^{(0)}$ as in (4.48).
for $t = 1 : T$ **do**
    for $i = 1 : Z_p$ **do**
        Update the velocity $\mathbf{V}_i^{(t)}$ as in (4.46).
        Update the position $\mathbf{X}_i^{(t)}$ as in (4.45).
        Find the personal best $\mathbf{X}_{\text{best},i}^{(t)}$ as in (4.47).
    end
    Find the global best $\mathbf{X}_{\text{best}}^{(t)}$ as in (4.48).
end
$\mathbf{x}_o = \mathbf{X}_{\text{best}}^{(T)}$
Update $\mathbf{B}_b, \mathbf{B}_{u,t}, \mathbf{B}_{u,r}$ for $\mathbf{x}_o$.

Then, the personal and global best solutions for ith particle during tth iteration are obtained as follows:

$$\mathbf{X}_{\text{best},i}^{(t)} = \underset{\mathbf{X}_i^{(t^*)}, \forall t^*=0,1,\cdots,t}{\arg\max} \quad R_T(\mathbf{F}_b, \mathbf{B}_b, \mathbf{F}_{u,t}, \mathbf{B}_{u,t}, \mathbf{F}_{u,r}, \mathbf{B}_{u,r}, \mathbf{P}, \mathbf{X}_i^{(t^*)}), \tag{4.47}$$

$$\mathbf{X}_{best}^{(t)} = \underset{\mathbf{X}_{\text{best},i}^{(t)}, \forall i=0,1,\cdots,Z_p}{\arg\max} \quad R_T(\mathbf{F}_b, \mathbf{B}_b, \mathbf{F}_{u,t}, \mathbf{B}_{u,t}, \mathbf{F}_{u,r}, \mathbf{B}_{u,r}, \mathbf{P}, \mathbf{X}_{\text{best},i}^{(t)}). \tag{4.48}$$

After T iterations, we update $\mathbf{x}_o = \mathbf{X}_{\text{best}}^{(T)}$. The pseudo-code is given in Algorithm 4.2, which summarizes the proposed PSO-based UAV positioning for equal PA with HBF.

4.3.3.3 Joint UAV Positioning and Optimal PA

In this problem, we jointly optimize UAV location $\mathbf{x}_u = [x_u, y_u]^T$ and $\mathbf{P}$, which are given by (4.35) and (4.44), respectively. To solve this non-convex optimization problem, we propose a PSO-based algorithmic solution to optimize $\mathbf{x}_u$ and $\mathbf{P}$, while

Algorithm 4.3: Proposed joint HBF, UAV location optimization and PA algorithm

Input: $Z_p, T, (\theta, \psi), (x_1, y_1, z_1), (x_u, y_u, z_u)$.
Output: $\mathbf{x}_o, \mathbf{P}, \mathbf{F}_b, \mathbf{B}_b, \mathbf{F}_{u,t}, \mathbf{B}_{u,t}, \mathbf{F}_{u,r}, \mathbf{B}_{u,r}$.
Formulate BS RF and BB stages using (4.20), (4.31).
Formulate UAV receive HBF stages using (4.22), (4.32).
Construct UAV transmit HBF stages via (4.29), (4.33).
for $i = 1 : Z_p$ **do**
  Initialize the velocity as $\mathbf{J}_{v,i}^{(0)} = \mathbf{0}$.
  Each entry of $\mathbf{J}_{p,i}^{(0)}$ is uniformly distributed in [0, 1].
  Set the personal best $\mathbf{J}_{p,\text{best},i}^{(0)} = \mathbf{J}_{p,i}^{(0)}$.
end
Find the global best $\mathbf{J}_{p,\text{best}}^{(0)}$ as in (4.53).
for $t = 1 : T$ **do**
  for $i = 1 : Z_p$ **do**
    Update the velocity $\mathbf{J}_{v,i}^{(t)}$ as in (4.51).
    Update the position $\mathbf{J}_{p,i}^{(t)}$ as in (4.50).
    Find the personal best $\mathbf{J}_{p,\text{best},i}^{(t)}$ as in (4.52).
  end
  Find the global best $\mathbf{J}_{p,\text{best}}^{(t)}$ as in (4.53).
end
$\mathbf{x}_o = \mathbf{J}_{p,\text{best}}^{(T)}, \mathbf{P} = \kappa_{\text{best}}^{(T)} \hat{\mathbf{P}}_{\text{best}}^{(T)}$
Update $\mathbf{B}_b, \mathbf{B}_{u,t}, \mathbf{B}_{u,r}$ for $\mathbf{x}_o$.

maximizing the total AR R_T. Here, the ith particle at the tth iteration now represents an instance of the UAV location and MU PA matrix, which is given as follows:

$$\mathbf{J}_{p,i}^{(t)} = [\mathbf{X}_i^{(t)}, \hat{\mathbf{P}}_i^{(t)}]^T = [x_i^{(t)}, y_i^{(t)}, \sqrt{\hat{p}_{1,i}^{(t)}}, \cdots, \sqrt{\hat{p}_{K,i}^{(t)}}]^T \in \mathbb{R}^{K+2}, \tag{4.49}$$

where each particle i represents the candidate UAV position and PA to K users and calculates the objective function as $R_T(\mathbf{F}_b, \mathbf{B}_b, \mathbf{F}_{u,t}, \mathbf{B}_{u,t}, \mathbf{F}_{u,r}, \mathbf{B}_{u,r}, \kappa_i^{(t)} \hat{\mathbf{P}}_i^{(t)}, \mathbf{X}_i^{(t)})$. Here, $\hat{\mathbf{P}}_i^{(t)} = \text{diag}(\sqrt{\hat{p}_{1,i}^{(t)}}, \cdots, \sqrt{\hat{p}_{K,i}^{(t)}}) \in \mathbb{R}^{K \times K}$ is the normalized PA matrix with $\hat{p}_{k,i}^{(t)} \in [0, 1]$, and similar to Sect. 4.3-C1, the transmit power constraints for any $\hat{\mathbf{P}}_i^{(t)}$ are satisfied by defining $\mathbf{P}_i^{(t)} = \kappa_i^{(t)} \hat{\mathbf{P}}_i^{(t)}$. Then, the position $\mathbf{J}_{p,i}^{(t)}$ and velocity $\mathbf{J}_{v,i}^{(t)}$ for ith particle during tth iteration are updated as follows:

$$\mathbf{J}_{p,i}^{(t+1)} = \mathbf{J}_{p,i}^{(t)} + \mathbf{J}_{v,i}^{(t+1)}, \tag{4.50}$$

$$\mathbf{J}_{v,i}^{(t+1)} = \gamma_1 \mathbf{Y}_1^{(t)} (\mathbf{J}_{p,\text{best}}^{(t)} - \mathbf{J}_{p,i}^{(t)}) + \gamma_2 \mathbf{Y}_2^{(t)} (\mathbf{J}_{p,\text{best},i}^{(t)} - \mathbf{J}_{p,i}^{(t)}) + \gamma_3^{(t)} \mathbf{J}_{v,i}^{(t)}. \tag{4.51}$$

Finally, the personal and global best solutions for ith particle during tth iteration are obtained as follows:

$$\mathbf{J}_{p,\text{best},i}^{(t)} = \underset{\mathbf{J}_{p,i}^{(t^*)},\forall t^*=0,1,\cdots,t}{\arg\max} \quad R_T(\mathbf{F}_b, \mathbf{B}_b, \mathbf{F}_{u,t}, \mathbf{B}_{u,t}, \mathbf{F}_{u,r}, \mathbf{B}_{u,r}, \kappa_i^{(t^*)}\hat{\mathbf{P}}_i^{(t^*)}, \mathbf{X}_i^{(t^*)}), \tag{4.52}$$

$$\mathbf{J}_{p,\text{best}}^{(t)} = \underset{\mathbf{J}_{p,\text{best},i}^{(t)},\forall i=0,1,\cdots,Z_p}{\arg\max} \quad R_T(\mathbf{F}_b, \mathbf{B}_b, \mathbf{F}_{u,t}, \mathbf{B}_{u,t}, \mathbf{F}_{u,r}, \mathbf{B}_{u,r}, \kappa_{\text{best},i}^{(t)}\hat{\mathbf{P}}_{\text{best,i}}^{(t)}, \mathbf{X}_i^{(t)}). \tag{4.53}$$

After T iterations, we update $\mathbf{x}_o = \mathbf{X}_{\text{best}}^{(T)}$ and $\mathbf{P} = \kappa_{\text{best}}^{(T)}\hat{\mathbf{P}}_{\text{best}}^{(T)}$. Algorithm 4.3 gives the pseudo-code of the proposed PSO-based joint UAV positioning and optimal PA with HBF.

4.4 Illustrative Results

In this section, we present the Monte Carlo simulation results based on the proposed algorithmic solutions. Table 4.1 outlines the simulation setup based on the 3-D microcell scenario [19] for the results discussed hereafter. Moreover, for PSO-based results, we use $Z_p = 20$, $\gamma_1 = \gamma_2 = 2$, and $\gamma_3 = 1.1$.

Table 4.1 Simulation parameters

Number of antennas		$(N_T, N_t, N_r) = 144$	
Number of paths	Path loss exponent	$L = 10$	3.6
Frequency	Channel bandwidth	28 GHz	100 MHz
Noise PSD	Reference path loss α	− 174 dBm/Hz	61.34 dB
BS/Gateway height	UAV height	10 m	20 m
User groups	# of users per group	$G = 1$ or 2	$K_g = \frac{K}{G}$
UAV x-axis range	UAV y-axis range	$[x_{\min}, x_{\max}] = [0, 100]$ m	$[y_{\min}, y_{\max}] = [0, 100]$ m
Mean AAoD/AAoA (1st link)	Mean AAoD (2nd link)	120°	$\phi_g = 21° + 120°(g-1)$
Mean EAoD/EAoA (1st link)	Mean EAoD (2nd link)	60°	$\theta_g = 60°$
Azimuth/elevation angle spread	# of network realizations	± 10°	2000
Minimum horizontal distance $\tau_{h,\min}$	Maximum horizontal distance $\tau_{h,\min}$	0	100 [m]

4.4.1 Spectral Efficiency

In this section, we compare the spectral efficiency of the proposed PSO-based algorithmic solutions for a UAV-assisted MU-mMIMO communications systems. To achieve this, we first analyze the rate of each link individually for a fixed UAV location and transmit power $P_T = 20$ dBm. Specifically, we consider that the BS is located at $(x_b, y_b, z_b) = (0, 0, 10)$, the UAV is deployed at fixed location $(x_u, y_u, z_u) = (50, 50, 20)$, and the users are randomly distributed and located at a far distance from BS $(x_k, y_k) \in [50, 100]$. We then compare the optimal UAV location for maximizing the capacity of the first link using two schemes: (1) exhaustive search; and (2) the proposed J-HBF-PSOL-EQPA over a given deployment span of $[(x_{\min}, y_{\min}), (x_{\max}, y_{\max})] = [0, 100]$. The results show that the optimal location for the UAV to maximize the capacity of the first link is close to the BS, as demonstrated in Fig. 4.3b. Furthermore, the proposed J-HBF-PSOL-EQPA can find the global optimal solution for almost 95% of the realizations, as shown in Fig. 4.3a. Figure 4.4 plots the rate of first link R_1 using J-HBF-PSOL-EQPA versus different 2-D UAV locations, which shows similar performance to exhaustive search solution presented in Fig. 4.3b.

Figure 4.5 analyzes the total rate of second link R_2 for exhaustive search and J-HBF-PSOL-EQPA. Due to the randomness in the placement of the users, finding a single optimal UAV location that covers all users, while minimizing interference, is a challenging task. However, the proposed J-HBF-PSOL-EQPA can find optimal UAV placement close to global solution as depicted in Fig. 4.5a. Figure 4.6 plots the AR of second link versus different UAV 2-D locations for $P_T = 20$ dBm, which shows that UAV placement by the proposed J-HBF-PSOL-EQPA can give higher AR when compared to UAV deployed at some fixed location. Figure 4.7 compares the AR versus transmit power P_T of the proposed HBF solution for four cases: (1)

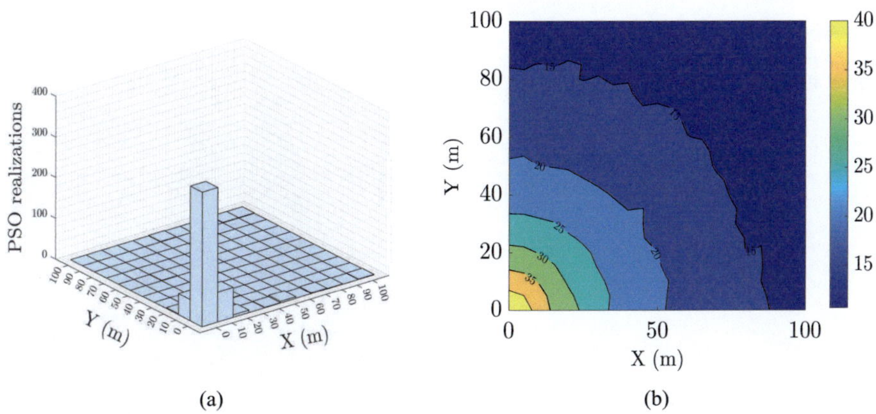

Fig. 4.3 Comparison of AR R_1. (**a**) Number of realizations for best UAV location using J-HBF-PSOL-EQPA. (**b**) Contour plot using exhaustive search

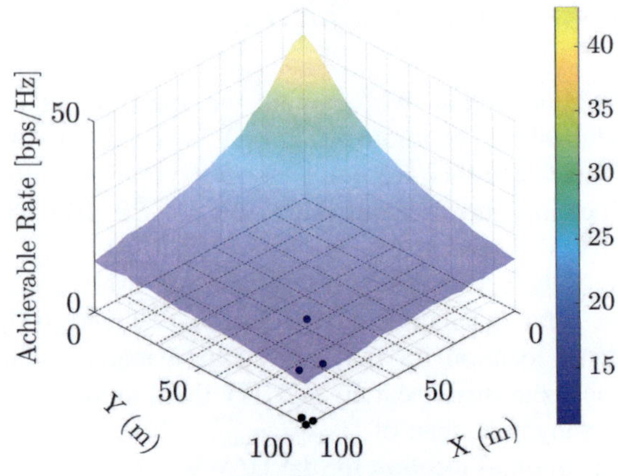

Fig. 4.4 AR R_1 versus $(x - y)$-coordinates at $P_T = 40$ dBm

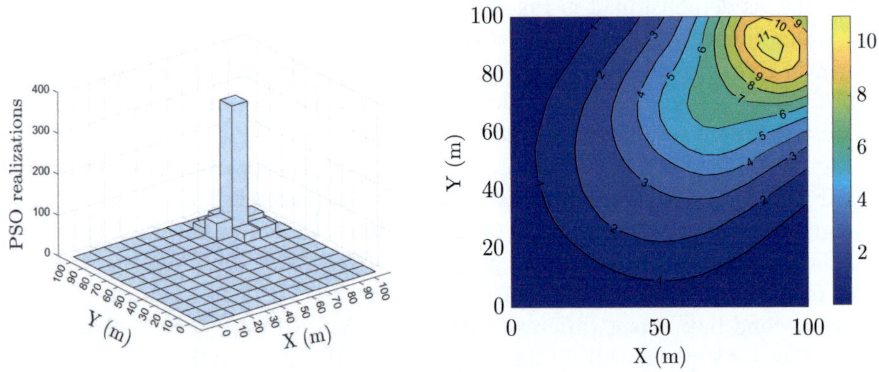

(a) (b)

Fig. 4.5 Comparison of AR R_2. (**a**) Number of realizations for best UAV location using J-HBF-PSOL-EQPA. (**b**) Contour plot using exhaustive search

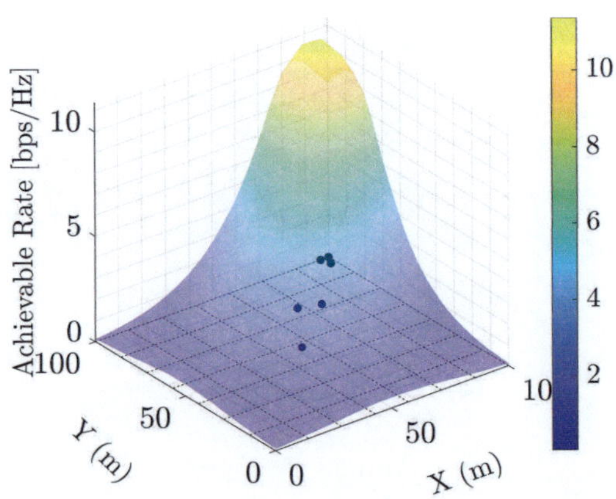

Fig. 4.6 AR R_2 versus $(x - y)$-coordinates at $P_T = 40$ dBm

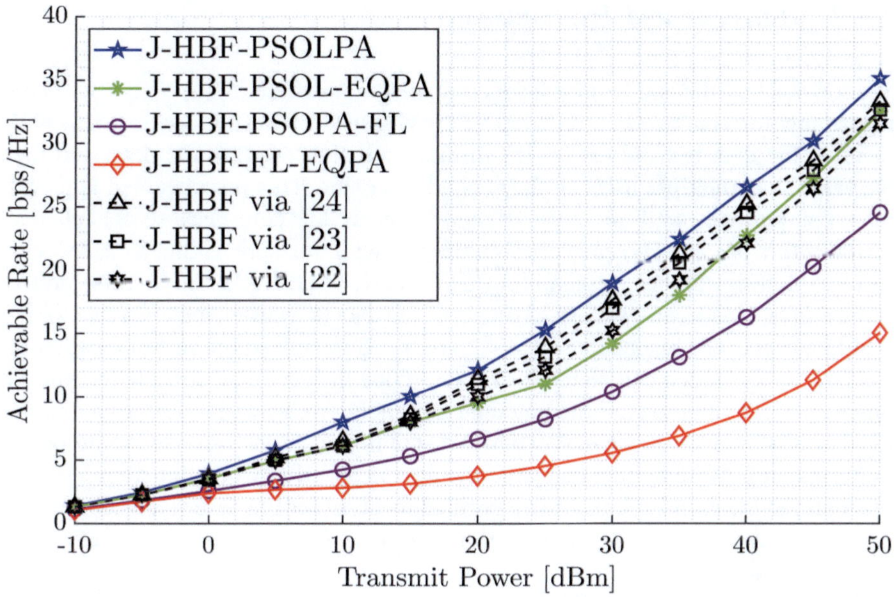

Fig. 4.7 Total AR R_T versus P_T for different proposed algorithmic schemes

J-HBF-PSOLPA; (2) J-HBF-PSOL-EQPA; (3) J-HBF-PSOPA-FL; and (4) J-HBF-FL-EQPA. The results show that all proposed algorithmic schemes can increase the total capacity of a UAV-assisted MU-mMIMO communications systems when compared to the J-HBF-FL-EQPA case. Comparing the single optimizations (i.e., PSO location-EQ PA and fixed location-PSO PA), we can see that optimizing UAV location can only provide a higher AR than allocating optimal power to multiple users. However, in a highly dynamic environment, where the users are far located and randomly distributed, the joint optimization of UAV location and PA can achieve higher spectral efficiency (i.e., $R_{PSOL-PSOPA} \geq R_{PSOL-EQPA} \gg R_{FL-PSOPA}$). Moreover, compared to J-HBF-FL-EQPA case at $P_T = 40$ dBm, the proposed J-HBF-PSOPA-FL, J-HBF-PSOL-EQPA, and J-HBF-PSOLPA schemes can increase the total AR by 77%, 155%, and 200%, respectively. We also compare the performance with the existing HBF solutions. For instance, compared to HBF schemes presented in [22, 23], the proposed J-HBF-PSOLPA can provide a higher AR. Similarly, compared to the iterative successive approximation (ISA) algorithmic solution in [24], which requires full CSI, the proposed J-HBF-PSOLPA can provide better performance with reduced CSI overhead size.

4.4.2 Buffer-Aided Transmission

Section 4.4.1 of this study presents a comparison of different proposed PSO-based algorithms for conventional relaying in UAV-assisted wireless systems, i.e., UAV receives the data transmitted by BS in first time slot, decodes the data, and then forwards it to multiple users in the second time slot. This prescheduled approach may not perform well in UAV-assisted wireless systems because the channel qualities (i.e., $\mathbf{H}_1$ and $\mathbf{H}_2$) can vary significantly with time, preventing the UAV relay from exploiting the best transmitting and receiving channels. In addition, unlike static relays, UAVs can fly close to BS, store the data in a buffer, and then fly close to multiple users. The results in Sect. 4.4.1 are based on a fixed optimized location for the UAV relay, which can increase the rate of one link but may degrade the rate for other link.[1] To fully exploit the potential of a mobile relay and to further maximize the capacity of UAV-assisted MU-mMIMO communications systems, we propose a buffer-aided UAV relay that can store the data in a buffer, while transitioning from one optimal location for link 1 to an optimal location for link 2. By considering two possible locations for UAV, we can maximize the SINR for each link, leading to a higher total rate.

In Fig. 4.8, we compare AR versus transmit power for UAVs with buffer and without buffer. For the buffer-aided UAV, we consider two scenarios for delay-unconstrained applications. In J-HBF-FL-EQPA, we consider UAV is deployed at fixed location ($x_u, y_u = 50, 50$), and multiple users have equal PA. However, the UAV does not transmit in consecutive time slots. In the second scenario, J-HBF-PSOLPA, we consider two different UAV locations (close to BS for link 1 and close to multiple users for link 2) and do not use prescheduled transmissions. The results show that using buffers at UAV can significantly increase the capacity (e.g., from 10 bps/Hz for J-HBF-FL-EQPA (without buffer) to 62 bps/Hz for J-HBF-PSOLPA (with buffer) at $P_T = 40$ dBm, which represents more than five-fold increase in capacity). It is important to note that we assume a buffer size as $B = \infty$, and thus, it represents a performance upper bound for a UAV-assisted MU-mMIMO system in delay-constrained transmissions. For delay-unconstrained transmissions, the average delay tends to ∞ as $B \rightarrow \infty$. However, with the simple heuristic modifications proposed in this subsection, the proposed algorithmic solutions for delay-unconstrained transmission can also be employed for delay-constrained transmission at the expense of a small performance degradation due to the delay constraint. Let D denote the waiting time (delay) experienced for a bit transmitted from the BS to the multiple users via UAV relay equipped with buffer and having queuing size Q bits. Then, for the HD UAV-assisted MU-mMIMO systems, we use Little's law to relate D to Q. Let $\lambda = \min(\mathrm{R}_1, \mathrm{R}_2)$ be the arrival

[1] Since R_2 is the minimum of $(\mathrm{R}_1, \mathrm{R}_2)$, UAV is deployed close to multiple users to maximize R_2; however, it increases the path loss for first link, which results in slight rate degradation for R_1 (Figs. 4.3, 4.4, 4.5, and 4.6).

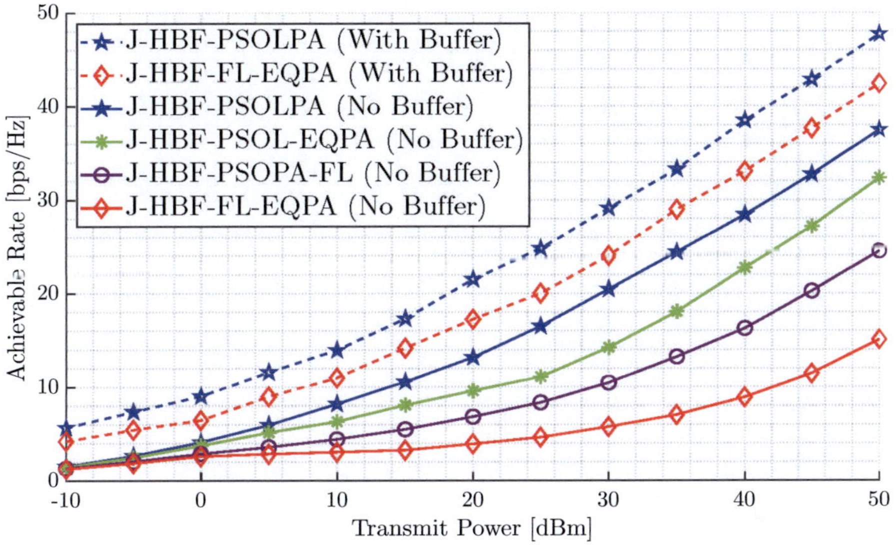

Fig. 4.8 Total AR R_T versus P_T for buffer-aided transmissions

rate, and then we can express D as [25]

$$D = \frac{Q}{\min(R_1, R_2)}. \tag{4.54}$$

We assume a first-in-first-out (FIFO) queuing mechanism for the UAV buffer, where the UAV receives data from the BS at a rate of R_1 and transmits it to the users at rate of R_2 incurring a delay D due to queuing bits Q of buffer.[2] Figure 4.9 plots the average delay versus transmit power P_T for the cases of: (1) J-HBF-FL-EQPA; and (2) J-HBF-PSOLPA. It can be seen that for a fixed queuing size of $Q = 2$ bits, increasing P_T can result in reduced average delay due to an increased AR for both cases. However, the proposed J-HBF-PSOLPA can reduce the average delay by approximately 50% when compared to J-HBF-FL-EQPA. In Fig. 4.10, we plot average delay versus queuing size Q for a fixed transmit power $P_T = 20$ dBm. As expected, higher queuing size results in increased delay. However, by utilizing the proposed J-HBF-PSOLPA, we can decrease D by more than 140% (e.g., at $Q = 8$ bits, D can be reduced from 0.48s to 0.2s). Finally, we compare the average delay versus P_T and Q for J-HBF-FL-EQPA and J-HBF-PSOLPA in Fig. 4.11. Our results show that the proposed J-HBF-PSOLPA can significantly reduce the average

[2] It is important to note that the arrival rate R_1 is greater than the departure rate R_2 for the given system (i.e., $R_1 > R_2$), and the stability of the system is still guaranteed due to the assumption of $\min(R_1, R_2)$ being used as the arrival rate. This ensures that the overall arrival rate is always less than or equal to the departure rate, which is necessary for the stability of the system.

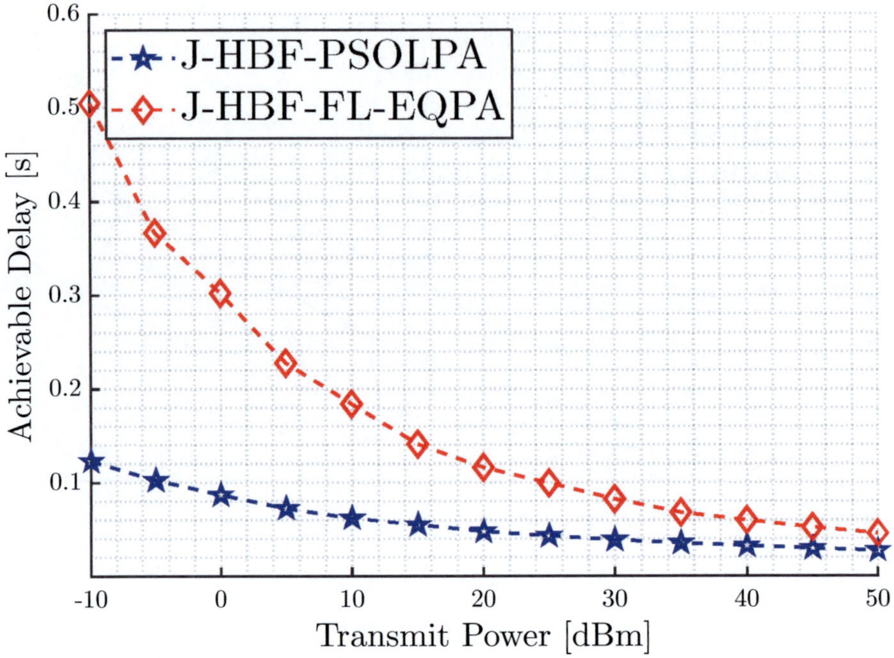

Fig. 4.9 Average delay D versus transmit power P_T

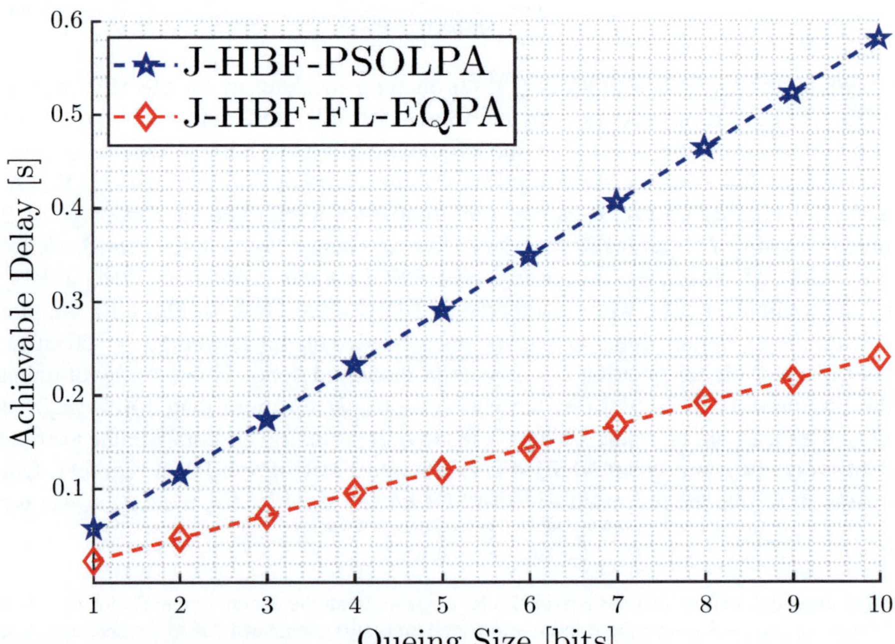

Fig. 4.10 Average delay D versus queuing size Q

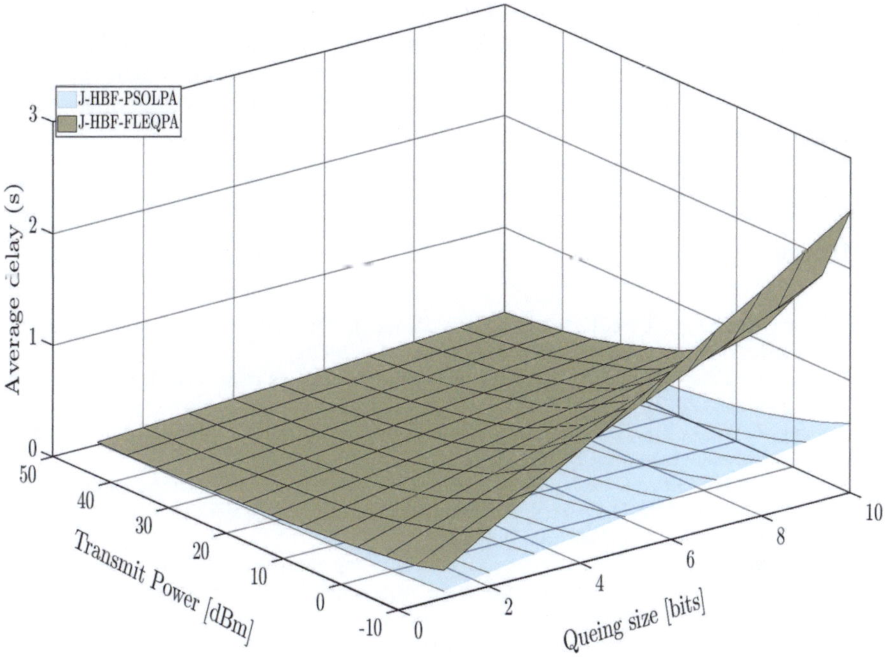

Fig. 4.11 Average delay D versus transmit power P_T and queuing size Q

delay for any combination of P_T and Q. Thus, it can be applied to both delay-unconstrained and delay-constrained applications in UAV-assisted MU-mMIMO communications systems.

4.5 Concluding Remarks

In this chapter, a UAV-assisted MU-mMIMO communications system has been considered, where the UAV operates as an DF relay between a BS and multiple users/devices. The problem of jointly designing the HBF stages for BS and UAV together with UAV deployment and optimal PA to multiple users is taken into consideration. For this challenging non-convex problem, we have proposed three different PSO-based algorithmic solutions to optimize the UAV location, and the power allocated to users, which are not directly accessible to the BS. Then, based on the optimized UAV location, and PA, the HBF stages for BS and UAV transmit and receive are sequentially updated for a maximum total AR. In particular, the RF stages are designed using the angular location of nodes, reducing the number of RF chains, while the BB stages are designed using reduced-dimension effective channel, which significantly reduces MU interference among the users. The results illustrate that the proposed PSO-based solutions can significantly enhance the capacity of a UAV-assisted MU-mMIMO communications system as well as reduce the average delay for delay-constrained transmissions, where the J-HBF-PSOLPA shows the superior performance by increasing the AR up to 200% compared to J-HBF-FL-EQPA.

References

1. S. Zhang, H. Zhang, Q. He, K. Bian, L. Song, Joint trajectory and power optimization for UAV relay networks. IEEE Commun. Lett. **22**(1), 161–164 (2018)
2. Y. Chen, W. Feng, G. Zheng, Optimum placement of UAV as relays. IEEE Commun. Lett. **22**(2), 248–251 (2017)
3. X. Jiang, Z. Wu, Z. Yin, Z. Yang, Power and trajectory optimization for UAV-enabled Amplify-and-Forward relay networks. IEEE Access **6**, 48688–48696 (2018)
4. M. Mahmood, A. Koc, T. Le-Ngoc, Spherical array-based joint beamforming and UAV positioning in massive MIMO systems, in *Proceedings of 2023 IEEE 97th Vehicular Technology Conference (VTC2023-Spring)* (2023), pp. 1–5
5. A. Koc, M. Mahmood, T. Le-Ngoc, Efficient dual-hop massive MIMO IoT networks with UAV DF relaying and hybrid beamforming, in *Proceedings of GLOBECOM 2023–2023 IEEE Global Communications Conference* (2023), pp. 6414–6419
6. L. Zhu, J. Zhang, Z. Xiao, X.-G. Xia, R. Zhang, Multi-UAV aided millimeter-wave networks: positioning, clustering, and beamforming. IEEE Trans. Wirel. Commun. **21**(7), 4637–4653 (2022)
7. J. Du, W. Xu, Y. Deng, A. Nallanathan, L. Vandendorpe, Energy-saving UAV-assisted multiuser communications with massive MIMO hybrid beamforming. IEEE Commun. Lett. **24**(5), 1100–1104 (2020)
8. W. Wang, N. Qi, L. Jia, C. Li, T. A. Tsiftsis, and M. Wang, Energy-efficient UAV-relaying 5G/6G spectrum sharing networks: interference coordination with power management and trajectory design. IEEE Open J. Commun. Soc. **3**, 1672–1687 (2022)
9. L. Sboui, H. Ghazzai, Z. Rezki, M.-S. Alouini, Achievable rates of UAV-relayed cooperative cognitive radio MIMO systems. IEEE Access **5**, 5190–5204 (2017)
10. F. Ono, H. Ochiai, R. Miura, A wireless relay network based on unmanned aircraft system with rate optimization. IEEE Trans. Wireless Commun. **15**(11), 7699–7708 (2016)
11. Y. Guo, S. Yin, J. Hao, Joint placement and resources optimization for multi-user UAV-relaying systems with underlaid cellular networks. IEEE Trans. Veh. Technol. **69**(10), 12374–12377 (2020)
12. X. Pang, J. Tang, N. Zhao, X. Zhang, Y. Qian, Energy-efficient design for mmWave-enabled NOMA-UAV networks. Sci. China Inf. Sci. **64**, 1–14 (2021)
13. W. Feng et al., Hybrid beamforming design and resource allocation for UAV-aided wireless-powered mobile edge computing networks with NOMA. IEEE J. Sel. Areas Commun. **39**(11), 3271–3286 (2021)
14. K. Liu, Y. Liu, P. Yi, Z. Xiao, X.-G. Xia, Deployment and robust hybrid beamforming for UAV MmWave communications. IEEE Trans. Commun. **71**(5), 3073–3086 (2023)
15. X. Yu, X. Huang, K. Wang, F. Shu, X. Dang, Joint design of power allocation, beamforming, and positioning for energy-efficient UAV-aided multiuser millimeter-wave systems. IEEE J. Sel. Areas Commun. **40**(10), 2930–2945 (2022)
16. W. Feng et al., Joint 3D trajectory and power optimization for UAV-aided mmWave MIMO-NOMA networks. IEEE Trans. Commun. **69**(4), 2346–2358 (2021)
17. Z. Xiao, H. Dong, L. Bai, D.O. Wu, X.-G. Xia, Unmanned aerial vehicle base station (UAV-BS) deployment with millimeter-wave beamforming. IEEE Internet Things J. **7**(2), 1336–1349 (2020)

18. R. Méndez-Rial, C. Rusu, N. González-Prelcic, A. Alkhateeb, R.W. Heath, Hybrid MIMO architectures for millimeter wave communications: phase shifters or switches? IEEE Access **4**, 247–267 (2016)
19. A. Koc, A. Masmoudi, T. Le-Ngoc, 3D angular-based hybrid precoding and user grouping for uniform rectangular arrays in massive MU-MIMO systems. IEEE Access **8**, 84689–84712 (2020)
20. M. Mahmood, A. Koc, T. Le-Ngoc, Energy-efficient MU-massive-MIMO hybrid precoder design: low-resolution phase shifters and digital-to-analog con- verters for 2D antenna array structures. IEEE Open J. Commun. Soc. **2**, 1842–1861 (2021)
21. M. Mahmood, A. Koc, T. Le-Ngoc, 3-D antenna array structures for millimeter wave multi user massive MIMO hybrid precoder design: a performance comparison. IEEE Commun. Lett. **26**(6), 1393–1397 (2022)
22. F. Sohrabi, W. Yu, Hybrid digital and analog beamforming design for large-scale antenna arrays. IEEE J. Sel. Topics Signal Process. **10**(3), 501–513 (2016)
23. X. Yu, J.-C. Shen, J. Zhang, K.B. Letaief, Alternating minimization algorithms for hybrid precoding in millimeter wave MIMO systems. IEEE J. Sel. Topics Signal Process. **10**(3), 485–500 (2016)
24. X. Xue, Y. Wang, L. Dai, C. Masouros, Relay hybrid precoding design in millimeter-wave massive MIMO systems. IEEE Trans. Signal Process. **66**(8), 2011–2026 (2018)
25. J.D. Little, A proof for the queuing formula: L = λw. Oper. Res. **9**(3), 383–387 (1961)

Chapter 5
Multiple UAV-Assisted Cooperative DF Relaying for Enhanced Coverage and Capacity

5.1 Introduction

In the previous chapter, we consider a single UAV operating as a relay between BS and obscured users, which can provide only limited user coverage and access. On the other hand, a network of multiple UAVs can efficiently enlarge the coverage region and increase the number of served users [1–3]. The authors in [1] propose a multi-UAV-relaying system, and compared the performance of a single multi-hop link and multiple dual-hop links. In [2], the position of multiple UAV-mounted BS is optimized to enhance the coverage area, while satisfying QoS requirements. Similarly, the placement and the number of UAVs are optimized in [3], while adhering to network capacity and coverage constraints. Research works such as [1–7] have ignored the direct link between the BS and Internet of Things (IoT) users. However, it is well known that when direct link is non-negligible or not too weak, the spatial diversity can be enhanced via direct and cooperative multiple paths (through UAV relays). In practice, assuming no direct link simplifies the design of the joint source-relay beamforming.

In this chapter, we consider a more practical cooperative transmission approach, integrating both the direct link from BS to IoT devices and the indirect links via multiple UAV relays to IoT users in MU-mMIMO IoT systems to enhance the capacity and overcome the coverage issues. Our objectives here are twofold: first, to show that the use of multiple UAVs as relays can significantly increase the sum rate capacity; and second, the optimization of UAV location, user association, and PA at BS and UAVs combined with HBF design can provide better performance than fixed UAV locations with equal PA.

T. Le-Ngoc et al., *AI-Enabled UAV-Assisted Massive MIMO*, Wireless Networks,
https://doi.org/10.1007/978-3-032-07872-8_5

The main contributions of this work are summarized as follows:

- We consider multiple UAVs-assisted cooperative DF relaying MU-mMIMO system, where the users received the signals from both BS and via UAV. When the direct link is involved, the received signals via source-to-destination link and multiple UAV relay-to-destination links are combined at the destination (user terminal) to enhance the overall signal strength. As such, the beamforming vector design at the BS needs to take into account both of these links. Furthermore, the optimization problem becomes more complex, which is solved using NI optimization solutions.
- The joint optimization of multiple UAV placements, PA at the BS, and the design of HBF for both the BS and UAVs are considered in this chapter. To solve the challenging non-convex problem, we utilize structured sequential optimization to address the multifaceted optimization problem by splitting it into two subproblems. First, K-means-based user clustering is used for UAV-users association based on the 3-D geometry-based mmWave channel model. Then, the location of each UAV is optimized jointly with PA using swarm intelligence. The RF beamforming stages for BS and UAVs are designed based on the slow time-varying AoD/AoA information, and BB stages are formulated using the reduced-dimensional effective channel matrices.
- The illustrative results show that multiple UAV-assisted cooperative relaying systems outperform a single UAV system in practical user distributions. Moreover, compared to fixed positions and equal PA of UAVs and BS, the joint optimization of UAV location and PA substantially enhances the total AR.

The rest of this chapter is organized as follows. Section 5.2 presents the system and channel model of multiple UAV-assisted MU-mMIMO system. In Sect. 5.3, we present the HBF design jointly with PA and UAV deployment. The illustrative results are provided in Sect. 5.4. Finally, the chapter is concluded in Sect. 5.5.

5.2 System and Channel Model

In this section, we introduce the system and channel models of the proposed multi-UAV-assisted relaying and the HBF design for a multiple dual-hop MU-mMIMO IoT system.

5.2.1 System Model

We consider a downlink MU-mMIMO IoT network, where a large number of nonoverlapping IoT devices are connected to an IoT gateway via wire or wireless links. Due to severe shadowing and blocking effect, many ground IoT devices experience low signal quality from the BS/eNodeB that is equipped with a large

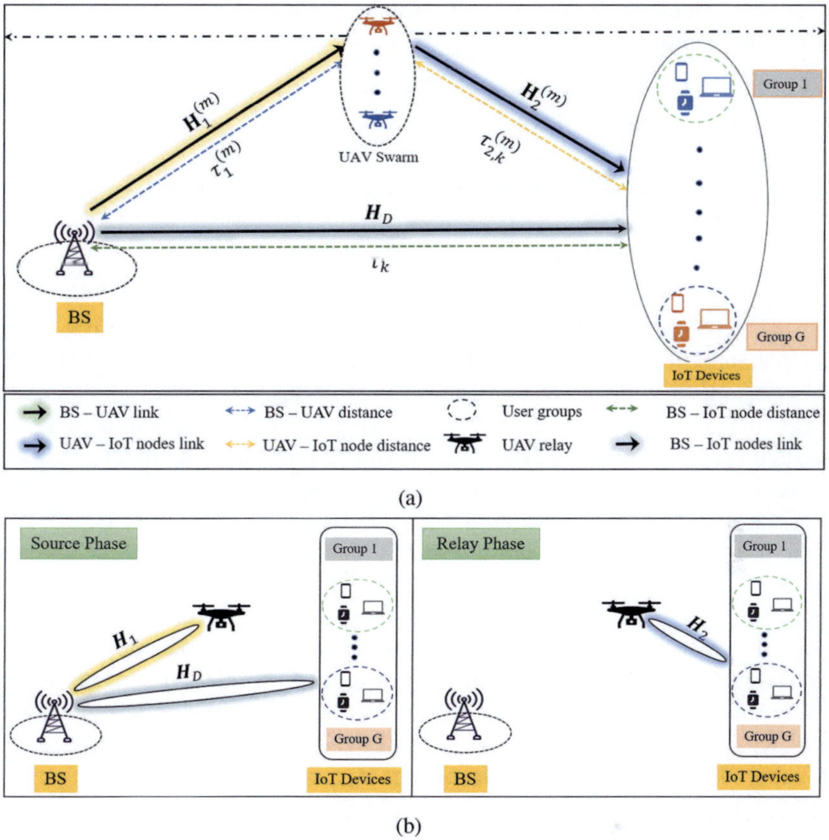

Fig. 5.1 Multiple UAV-assisted MU-mMIMO IoT communications. (**a**) Network model. (**b**) UAV as DF relay transmission phase

array having N_T antenna elements. To address this challenging scenario, we consider a cooperative relaying system model as shown in Fig. 5.1a, to serve $\mathbb{K} = \{1, \cdots, K\}$ single-antenna IoT nodes, which are clustered in G groups, where gth group has K_g IoT nodes such that $K = \sum_{g=1}^{G} K_g$. Then, M_u different UAVs, indexed by the set $\mathbb{U} = \{u_1, u_2, \cdots, u_{M_u}\}$, are deployed to serve $K = M_u K_m$ IoT users, where K_m is the number of users served by mth UAV, which operates as DF relay between BS/eNodeB and IoT node.[1] Let (x_b, y_b, z_b), $(x_u^{(m)}, y_u^{(m)}, z_u^{(m)})$ and (x_k, y_k, z_k) denote the locations of BS, mth UAV relay, and kth IoT user,

[1] We assume equal UAV-user clustering for simplicity. However, it can be applicable for unequal user clustering, which is left as our future work.

respectively. Then, we define the 3-D distances for multiple UAV-assisted MU-mMIMO IoT system as follows:

$$\begin{aligned} \tau_1^{(m)} &= \sqrt{(x_u^{(m)} - x_b)^2 + (y_u^{(m)} - y_b)^2 + (z_u^{(m)} - z_b)^2} \\ \tau_{2,k}^{(m)} &= \sqrt{(x_u^{(m)} - x_k)^2 + (y_u^{(m)} - y_k)^2 + (z_u^{(m)} - z_k)^2} \\ \tau_k &= \sqrt{(x_b - x_k)^2 + (y_b - y_k)^2 + (z_b - z_k)^2}, \end{aligned} \tag{5.1}$$

where $\tau_1^{(m)}$, $\tau_{2_k}^{(m)}$, and τ_k are the 3-D distance between mth UAV and BS, between mth UAV and kth IoT node, and between BS and kth IoT node, respectively. Each UAV is equipped with $N_r (N_t)$ antennas for receiving (transmitting) signals from (to) BS (IoT users). For simplicity, we assume a homogeneous fleet of UAVs with consistent specifications and functionality. Unlike traditional static relaying, which uses fixed relay locations, we presumptively use multiple UAVs as movable relays.

For the downlink transmission of $N_s = K$ data streams, we consider HBF for BS and all UAV relays as shown in Fig. 5.2. The BS consists of RF beamforming stage $\mathbf{F}_b \in \mathbb{C}^{N_T \times N_{RF_b}}$, BB stage $\mathbf{B}_b \in \mathbb{C}^{N_{RF_b} \times K}$, and MU PA matrix $\mathbf{P}_b = \text{diag}(\sqrt{p_{b_1}}, \cdots, \sqrt{p_{b_K}}) \in \mathbb{C}^{K \times K}$. Here, N_{RF_b} is the RF chains such that $N_s \leq N_{RF_b} \leq N_T$ to guarantee multistream transmission. We consider half-duplex (HD) DF relaying for each UAV. Therefore, each round of information transmission from BS to IoT nodes can be divided into two phases: (1) source phase (SP) and (2) relay phase (RP), as illustrated in Fig. 5.1b. In SP, BS transmits K data streams to the following: (1) K IoT users through channel $\mathbf{H}_D \in \mathbb{C}^{K \times N_T}$ and (2) each mth UAV via channel $\mathbf{H}_1^{(m)} \in \mathbb{C}^{N_r \times N_T}$. Using N_r antennas, each UAV receives signals with RF stage $\mathbf{F}_{u,r}^{(m)} \in \mathbb{C}^{N_{RF_u} \times N_r}$ and BB stage $\mathbf{B}_{u,r}^{(m)} \in \mathbb{C}^{K \times N_{RF_u}}$. In RP, each UAV decodes the received information and then forwards the decoded information to K IoT nodes using RF beamformer $\mathbf{F}_{u,t}^{(m)} \in \mathbb{C}^{N_t \times N_{RF_u}}$, BB stage $\mathbf{B}_{u,t}^{(m)} \in \mathbb{C}^{N_{RF_u} \times K}$, and MU PA matrix $\mathbf{P}_u^{(m)} = \text{diag}(\sqrt{p_{u_1}^{(m)}}, \cdots, \sqrt{p_{u_K}^{(m)}}) \in \mathbb{C}^{K \times K}$ through channel $\mathbf{H}_2^{(m)}$

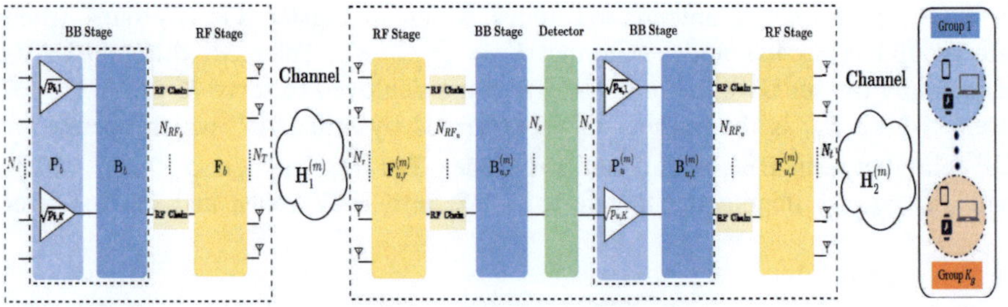

Fig. 5.2 Multiple UAV-assisted MU-mMIMO HBF system model

$\in \mathbb{C}^{K\times N_t}$, where $p_{u_k}^{(m)}$ reflects the allocated power to kth user from mth UAV. The implementation of all RF beamforming/combining stages involves the use of PSs and thus imposes a CM constraint, i.e., $|\mathbf{F}_b(i,j)| = \frac{1}{\sqrt{N_T}}$, $|\mathbf{F}_{u,r}^{(m)}(i,j)| = \frac{1}{\sqrt{N_r}}$, $|\mathbf{F}_{u,t}^{(m)}(i,j)| = \frac{1}{\sqrt{N_t}}$ $\forall i, j, m$. For the data signal $\mathbf{d} = [d_1, d_2, \cdots, d_K]^T$ with $\mathbb{E}\{\mathbf{d}\mathbf{d}^H\} = \mathbf{I}_K \in \mathbb{C}^{K\times K}$, the signal received at mth UAV (after BB processing) during SP is given as follows:

$$\tilde{\mathbf{y}}_u^{(m)} = \mathbf{B}_{u,r}^{(m)}\mathbf{F}_{u,r}^{(m)}\mathbf{H}_1^{(m)}\mathbf{F}_b\mathbf{B}_b\mathbf{P}_b\mathbf{d} + \mathbf{B}_{u,r}^{(m)}\mathbf{F}_{u,r}^{(m)}\mathbf{n}_u^{(m)}, \tag{5.2}$$

where $n_u^{(m)} \in \mathbb{C}^{N_{u,r}}$ denotes the zero-mean complex circularly symmetric Gaussian noise vector at mth UAV relay with covariance matrix $\mathbb{E}\{\mathbf{n}_u\mathbf{n}_u^H\} = \sigma_{n_u}^2\mathbf{I}_{N_{u,r}} \in \mathbb{C}^{N_{u,r}\times N_{u,r}}$. Then, the signal transmitted by mth UAV during RP is given as

$$\hat{\mathbf{s}}^{(m)} = \mathbf{F}_{u,t}^{(m)}\mathbf{B}_{u,t}^{(m)}\mathbf{P}_u^{(m)}\hat{\mathbf{d}}^{(m)}, \tag{5.3}$$

where $\hat{\mathbf{d}}^{(m)}$ is the re-encoded signal at mth UAV relay. Each IoT node receives signal from BS and each mth UAV during SP and RP, respectively. Then, the received signal at kth IoT node from BS and mth UAV can be written as

$$\begin{aligned} y_k^{(m)} &= y_{k,SP} + y_{k,RP}^{(m)}, \\ &= \mathbf{h}_{D,k}^T\mathbf{F}_b\mathbf{B}_b\mathbf{P}_b\hat{d}_k + n_{D_k} + \mathbf{h}_{2,k}^{T(m)}\mathbf{F}_{u,t}^{(m)}\mathbf{B}_{u,t}^{(m)}\mathbf{P}_u^{(m)}\hat{d}_k^{(m)} + n_{2_k}^{(m)}, \\ &= \underbrace{\sqrt{p_{b_k}}\mathbf{h}_{D,k}^T\mathbf{F}_b\mathbf{B}_b\mathbf{P}_b\hat{d}_k + \sqrt{p_{u_k^{(m)}}}\mathbf{h}_{2,k}^{T(m)}\mathbf{F}_{u,t}^{(m)}\mathbf{B}_{u,t}^{(m)}\mathbf{P}_u^{(m)}\hat{d}_k^{(m)}}_{\text{Desired Signal}} + \\ &\underbrace{\sum_{\hat{k}=1}^{K}\sqrt{p_{b_{\hat{k}}}}\mathbf{h}_{D,k}^T\mathbf{F}_b\mathbf{B}_b\mathbf{P}_b\hat{d}_{\hat{k}} + \sum_{\hat{k}=1}^{K}\sqrt{p_{u_k^{(m)}}}\mathbf{h}_{2,k}^{T(m)}\mathbf{F}_{u,t}^{(m)}\mathbf{B}_{u,t}^{(m)}\mathbf{P}_u^{(m)}\hat{d}_{\hat{k}}^{(m)}}_{\text{Total MU-interference}} \\ &+ \underbrace{n_{D_k} + n_{2_k}^{(m)}}_{\text{Total Noise}}, \end{aligned} \tag{5.4}$$

where $n_{D_k} \sim \mathcal{CN}(\mathbf{0}, \sigma_{n_D}^2)$ and $n_{2_k}^{(m)} \sim \mathcal{CN}(\mathbf{0}, \sigma_{n_2}^2)$ are the additive circular symmetric Gaussian noise at kth IoT node. The power constraint of the beamforming matrices can be expressed as $\|\mathbf{F}_b\mathbf{B}_b\mathbf{P}_b\|_F^2 = P_T$ and $\|\mathbf{F}_{u,t}^{(m)}\mathbf{B}_{u,t}^{(m)}\mathbf{P}_u^{(m)}\|_F^2 = P_u^{(m)}$, where P_T and $P_u^{(m)}$ denote the total transmit power of BS and mth UAV, respectively. Then, the AR of the first link (i.e., BS $\rightarrow$ mth UAV) is given as follows:

$$\mathrm{R}_1^{(m)} = \log_2\left|\mathbf{I}_K + \mathbf{Q}_1^{-1(m)}\mathbf{B}_{u,r}^{(m)}\mathcal{H}_1^{(m)}\mathbf{B}_b\mathbf{B}_b^H\mathcal{H}_1^{H(m)}\mathbf{B}_{u,r}^{H(m)}\right|, \tag{5.5}$$

where $\mathbf{Q}_1^{-1(m)} = (\sigma_{n_u}^2 \mathbf{B}_{u,r}^{(m)} \mathbf{F}_{u,r}^{(m)})^{-1} \mathbf{F}_{u,r}^{H(m)} \mathbf{B}_{u,r}^{H(m)}$ and $\mathcal{H}_1^{(m)} = \mathbf{F}_{u,r}^{(m)} \mathbf{H}_1^{(m)} \mathbf{F}_b$. The SINR of kth IoT node via mth UAV is given as [8]

$$\mathrm{SINR}_k^{(m)} = \frac{p_{b_k} |\mathbf{h}_{D_k}^H \mathbf{F}_b \mathbf{b}_{b_k}|^2}{\sum\limits_{\hat{k} \neq k} p_{b_{\hat{k}}} |\mathbf{h}_{D_k}^H \mathbf{F}_b \mathbf{b}_{b_{\hat{k}}}|^2 + \sigma_{nD}^2} + \frac{p_{u_k}^{(m)} |\mathbf{h}_{2_k}^{H(m)} \mathbf{F}_{u,t}^{(m)} \mathbf{b}_{u,t_k}^{(m)}|^2}{\sum\limits_{\hat{k} \neq k} p_{u_{\hat{k}}}^{(m)} |\mathbf{h}_{2_k}^{H(m)} \mathbf{F}_{u,t}^{(m)} \mathbf{b}_{u,t_{\hat{k}}}^{(m)}|^2 + \sigma_{n_2}^{2(m)}}. \tag{5.6}$$

The AR for the mth second link (i.e., from mth UAV to K_m users) can be written as

$$\mathrm{R}_2^{(m)} = \mathbb{E}\{ \sum_{i=1}^{K_m} \log_2(1 + \mathrm{SINR}_i^{(m)}) \}. \tag{5.7}$$

For multiple dual-hop MU-mMIMO IoT system, where each mth UAV is deployed at a fixed height z_u^m, and relaying data to K_m IoT nodes, the total AR can be maximized by the joint optimization of $\mathbf{F}_b$, $\mathbf{B}_b$, $\mathbf{F}_{u,t}^{(m)}$, $\mathbf{F}_{u,r}^{(m)}$, $\mathbf{B}_{u,t}^{(m)}$, $\mathbf{B}_{u,r}^{(m)}$, $\mathbf{P}_b$, $\mathbf{P}_u^{(m)}$ and the UAV location $\mathbf{x}_u^{(m)} = [x_u^{(m)}, y_u^{(m)}]^T \in \mathbb{R}^2$, which is to be optimized within the given deployment area. Then, we can formulate the optimization problem as

$$\begin{aligned}
&\max_{\left\{\mathbf{F}_b, \mathbf{B}_b, \mathbf{F}_{u,t}^{(m)}, \mathbf{B}_{u,t}^{(m)}, \mathbf{F}_{u,r}^{(m)}, \mathbf{B}_{u,r}^{(m)}, \mathbf{P}_b, \mathbf{P}_u^{(m)}, \mathbf{x}_u^{(m)}\right\}} \quad \mathrm{R}_T \\
&\text{s.t.} \quad C_1 : |\mathbf{F}_{u,t}^{(m)}(i,j)| = \frac{1}{\sqrt{N_t}}, |\mathbf{F}_{u,r}^{(m)}(i,j)| = \frac{1}{\sqrt{N_r}}, \forall i, j, m, \\
&\qquad C_2 : |\mathbf{F}_b(i,j)| = \frac{1}{\sqrt{N_T}}, \forall i, j, \\
&\qquad C_3 : \bigcup_{m \in M_u} K_m = K, \quad \forall m, \\
&\qquad C_4 : \mathbb{E}\{\|\mathbf{s}_1\|_2^2\} \le P_T, \mathbb{E}\{\|\hat{\mathbf{s}}^{(m)}\|_2^2\} \le P_T, \forall m, \\
&\qquad C_5 : p_{b_k} \ge 0, p_{u_k}^{(m)} \ge 0, \quad \forall k, m, \\
&\qquad C_6 : \mathbf{x}_{\min} \le \mathbf{x}_u^{(m)} \le \mathbf{x}_{\max}, \quad \forall m,
\end{aligned} \tag{5.8}$$

where $\mathrm{R}_T = \sum_{m=1}^{M_u} (1/2) \min(\mathrm{R}_1^{(m)}, \mathrm{R}_2^{(m)})$ is the total AR, C_1 and C_2 refer to the CM constraint due to the use of PSs for UAV and BS, respectively, C_3 ensures that the total user count is consistent with the counts for each UAV, C_4 indicates the total transmit power constraint for UAV and BS, C_5 represents the nonnegative allocated power to each IoT node from BS and each UAV, and C_6 implies UAV deployment within the given flying span. Here, $[\mathbf{x}_{\min}, \mathbf{x}_{\max}] = [(x_{\min}, y_{\min}), (x_{\max}, y_{\max})]$ represents the deployment range for each UAV in 2-D space. The optimization problem defined in (5.8) is non-convex and intractable. Thus, we develop a suboptimal solution for (5.8) in Sect. 5.3.

5.2.2 Channel Model

We consider mmWave channels for both links. The channel between BS and mth UAV is modeled based on the Saleh-Valenzuela channel model [9] and is given as

$$\mathbf{H}_1^{(m)} = \sum_{c=1}^{C}\sum_{l=1}^{L} z_{1_{cl}}^{(m)} \tau_{1_{cl}}^{-\eta(m)} \boldsymbol{\phi}_{1,r}^{(m)}(\theta_{r_{cl}}^{(m)}, \psi_{r_{cl}}^{(m)}) \boldsymbol{\phi}_{1,t}^{T(m)}(\theta_{t_{cl}}^{(m)}, \psi_{t_{cl}}^{(m)}), \tag{5.9}$$

where C is the total number of clusters, L is the total number of paths, η is the path loss exponent, $z_{1_{cl}}^{(m)}$ is the complex gain of lth path in cth cluster for mth UAV, and $\boldsymbol{\phi}_{1,j}^{(m)}(\cdot,\cdot)$ is the corresponding transmit or receive array steering vector for URA, which is given as [10]

$$\begin{aligned}\boldsymbol{\phi}_{1,j}^{(m)}(\theta^{(m)}, \psi^{(m)}) &= \left[1, \cdots, e^{-j2\pi d(N_x-1)\sin(\theta^{(m)})\cos(\psi^{(m)})}\right] \\ &\otimes \left[1, \cdots, e^{-j2\pi d(N_y-1)\sin(\theta^{(m)})\sin(\psi^{(m)})}\right],\end{aligned} \tag{5.10}$$

where $j = \{t, r\}$, d is the inter-element spacing, and $N_x (N_y)$ is the horizontal (vertical) size of corresponding antenna array at BS and UAV. Here, the angles $\theta_{j_{cl}}^{(m)} \in \left[\theta_{j_c}^{(m)} - \delta_{j_c}^{\theta(m)}, \theta_{j_c}^{(m)} + \delta_{j_c}^{\theta(m)}\right]$ and $\psi_{j_{cl}}^{(m)} \in \left[\psi_{j_c}^{(m)} - \delta_{j_c}^{\psi(m)}, \psi_{j_c}^{(m)} + \delta_{j_c}^{\psi(m)}\right]$ are the azimuth and elevation AoD ($j = t$) or AoA ($j = r$) for lth path in channel $\mathbf{H}_1^{(m)}$, respectively. Here, $\theta_{j_c}^{(m)}$ and $\psi_{j_c}^{(m)}$ are the mean azimuth and elevation angles, respectively, with $\delta_{j_c}^{\theta(m)}$ ($\delta_{j_c}^{\psi(m)}$) represents the azimuth (elevation) angle spread. The channel vector between the UAV (or BS) and the kth IoT node can be written as

$$\mathbf{h}_{i,k}^T = \sum\nolimits_{c=1}^{C} z_{i,k_c} \tau_{i,k_c}^{-\eta} \boldsymbol{\phi}(\theta_{k_c}, \psi_{k_c}) = \mathbf{z}_{i,k}^T \boldsymbol{\Phi}_{i,k} \in \mathbb{C}^N, \tag{5.11}$$

where $i = \{D, 2\}$, C is the total number of downlink paths from UAV (or BS) to IoT nodes, $z_{i,k_l} \sim \mathcal{CN}(0, \frac{1}{L})$ is the complex path gain of lth path, and $\boldsymbol{\phi}(\cdot,\cdot) \in \mathbb{C}^N$ is the downlink array phase response vector. Then, the complete channel matrix for K IoT nodes can be written as

$$\mathbf{H}_i = [\mathbf{h}_{i,1}, \cdots, \mathbf{h}_{i,K}]^T = \mathbf{Z}_i \boldsymbol{\Phi}_i \in \mathbb{C}^{K \times N_t}, \tag{5.12}$$

where $\mathbf{Z}_i = [\mathbf{z}_{i,1}, \cdots, \mathbf{z}_{i,K}]^T \in \mathbb{C}^{K \times L}$ is the complete path gain matrix for all downlink IoT nodes, and $\boldsymbol{\Phi}_{i,k} \in \mathbb{C}^{L \times N}$ is the slow time-varying array phase response matrix.

5.3 Proposed Joint User Association, Multiple UAV Positioning, PA and HBF

In this section, our objective is to optimize each UAV location jointly with PA from BS and UAV and sequentially design HBF stages for BS and each UAV to reduce the CSI overhead size, while maximizing the throughput of a multiple UAV-assisted MU-mMIMO IoT system. First, we discuss the UAV-user association using K-means-based user clustering.

5.3.1 UAV-User Association

The proposed scheme leverages K-means-based user association to assign K_m users to the nearest mth UAV, while maintaining exclusive user-UAV associations. The objective is to minimize the sum of squared distances between users and their assigned mth UAV (i.e., $\min \sum_{m=1}^{M} \sum_{k=1}^{K} z_k^{(m)} ||\mathbf{x}_k - \mathbf{x}_u^{(m)}||^2$). Here, $z_k^{(m)}$ is the assignment variable, which is defined as

$$z_k^{(m)} = \begin{cases} 1, & \text{if user } k \text{ is assigned to UAV } m, \\ 0, & \text{otherwise.} \end{cases} \tag{5.13}$$

5.3.2 Joint UAV Deployment, Optimal PA, and HBF Design

In this section, our objective is to design the HBF stages for BS and M_u UAVs by using sequential optimization. Initially, both RF and BB stages are constructed using some fixed UAV locations. Then, we employ swarm intelligence to optimize each UAV location as well as PA from BS and each UAV for maximum total AR. Finally, the RF and BB stages are reformulated for the optimal UAV location as well as the allocated power in the MU PA blocks $\mathbf{P}_b$ and $\mathbf{P}_u^{(m)}$.

5.3.2.1 RF and BB Stage Design

The RF beamforming stage for BS and each UAV (both transmit and receive) are designed as

$$\mathbf{F} = \left[\mathbf{e}_j\left(\lambda_x^{u_1}, \lambda_y^{k_1}\right), \cdots, \mathbf{e}_j\left(\lambda_x^{u_{N_{RF}}}, \lambda_y^{k_{N_{RF}}}\right)\right] \in \mathbb{C}^{N_T \times N_{RF}}, \tag{5.14}$$

where $j = \{t, r\}$ and $\mathbf{e}(\cdot, \cdot)$ is the corresponding transmit or receive steering vector, which is defined as

$$\mathbf{e}(\theta, \psi) = \frac{1}{\mathcal{N}_t}\left[1, e^{j2\pi d\sin(\theta)\cos(\psi)}, \cdots, e^{j2\pi d(\mathcal{N}_{x,t}-1)\sin(\theta)\cos(\psi)}\right]^T \otimes\left[1, e^{j2\pi d\sin(\theta)\sin(\psi)}, \cdots, e^{j2\pi d(\mathcal{N}_{y,t}-1)\sin(\theta)\sin(\psi)}\right]^T, \tag{5.15}$$

where $\mathcal{N}_T = \{N_T, N_t, N_r\}$. Here, the RF beamformers are constructed via quantized angle pairs, which are defined as $\lambda_x^u = -1 + \frac{2u-1}{\mathcal{N}_{x,t}}$ for $u = 1, \cdots, \mathcal{N}_{x,t}$ and $\lambda_y^k = -1 + \frac{2k-1}{\mathcal{N}_{y,t}}$ for $k = 1, \cdots, \mathcal{N}_{y,t}$. The quantized angle pairs reduce the number of RF chains at BS and each UAV, while providing complete AoD/AoA supports, which are defined as

$$\text{AoD} = \left\{\sin(\theta)\left[\cos(\psi), \sin(\psi)\right] \middle| \theta \in \boldsymbol{\theta}_t, \psi \in \boldsymbol{\psi}_t\right\}, \tag{5.16}$$

$$\text{AoA} = \left\{\sin(\theta)\left[\cos(\psi), \sin(\psi)\right] \middle| \theta \in \boldsymbol{\theta}_r, \psi \in \boldsymbol{\psi}_r\right\}, \tag{5.17}$$

where $\boldsymbol{\theta}_i = \left[\theta_i - \delta_i^\theta, \theta_i + \delta_i^\theta\right]$ and $\boldsymbol{\psi}_i = \left[\psi_i - \delta_i^\psi, \psi_i + \delta_i^\psi\right]$ denote the azimuth and elevation angle supports, respectively. After designing the transmit and receive RF beamformers for BS, and each UAV, the effective channel matrices $\mathcal{H}_1^{(m)}$ and $\mathcal{H}_2^{(m)}$ as seen from the BB stages are given as follows:

$$\mathcal{H}_1^{(m)} = \mathbf{F}_{u,r}^{(m)}\mathbf{H}_1^{(m)}\mathbf{F}_b = \mathcal{Q}_1^{(m)}\boldsymbol{\Sigma}_1^{(m)}\mathcal{V}_1^{H(m)}, \tag{5.18}$$

$$\mathcal{H}_2^{(m)} = \begin{bmatrix} \mathbf{H}_{2,1}^{(m)}\mathbf{F}_{u,t,1}^{(m)} & \cdots & \mathbf{H}_{2,1}^{(m)}\mathbf{F}_{u,t,G}^{(m)} \\ \vdots & \ddots & \vdots \\ \mathbf{H}_{2,G}^{(m)}\mathbf{F}_{u,t,1}^{(m)} & \cdots & \mathbf{H}_{2,G}^{(m)}\mathbf{F}_{u,t,G}^{(m)} \end{bmatrix}, \tag{5.19}$$

where $\mathcal{Q}_1^{(m)} \in \mathbb{C}^{N_{RF_u} \times rank(\mathcal{H}_1^{(m)})}$ and $\mathcal{V}_1^{(m)} \in \mathbb{C}^{N_{RF_b} \times rank(\mathcal{H}_1^{(m)})}$ are tall unitary matrices and $\boldsymbol{\Sigma}_1^{(m)}$ is the diagonal matrix with singular values in the decreasing order for mth UAV. Then, $\mathbf{B}_{u,r}^{(m)}$ for mth UAV is defined as

$$\mathbf{B}_{u,r}^{(m)} = \mathcal{Q}_1^{H(m)}. \tag{5.20}$$

The reduced-size effective CSI $\mathcal{H}_2^{(m)}$ given in (5.19) is employed for designing $\mathbf{B}_{u,t}^{(m)}$ by using the RZF technique and is defined as follows:

$$\mathbf{B}_{u,t}^{(m)} = (\mathcal{H}_2^{H(m)}\mathcal{H}_2^{(m)} + \xi^{(m)} N_{RF_u}\mathbf{I}_{N_{RF_u}})^{-1}\mathcal{H}_2^{H(m)}, \tag{5.21}$$

where $\xi^{(m)} = \frac{\sigma^{2(m)}}{P_T^{(m)}}$ is the regularization parameter and $\mathbf{I}_{N_{RF_u}} \in \mathbb{C}^{N_{RF_u} \times N_{RF_u}}$. The optimal design of $\mathbf{B}_b$ is formulated using an effective channel $\mathcal{H}_{1D}$, which constitutes all channel components from BS (i.e., $\mathbf{H}_D$ and $\mathbf{H}_2^{(m)}, \forall m = 1, \cdots, M$). Then, the total effective channel can be written as

$$\mathcal{H}_{1D} = [\mathbf{H}_D, \mathbf{H}_1^{(1)}, \mathbf{H}_1^{(2)}, \cdots, \mathbf{H}_1^{(M)}] \in \mathbb{C}^{(MN_r+K)\times K}. \tag{5.22}$$

By using SVD of effective channel $\mathcal{H}_{1D}$, we can design $\mathbf{B}_b$ using tall unitary matrix $\mathcal{V}_{1D} \in \mathbb{C}^{N_{RF_b} x rank(\mathcal{H}_{1D})}$ as [10]

$$\mathbf{B}_b = \sqrt{\frac{P_T}{K}} \mathcal{V}_{1D} \in \mathbb{C}^{N_{RF_b} \times K}. \tag{5.23}$$

5.3.2.2 Joint Multiple UAV Positioning and Optimal PA

After the design of RF and BB stages for BS and UAVs, the optimization problem given in (5.8) can be reformulated as

$$\begin{aligned} &\max_{\left\{\mathbf{P}_b, \mathbf{P}_u^{(m)}, \mathbf{x}_u^{(m)}\right\}} \quad \mathrm{R}_T \\ &C_4 : \mathbb{E}\{\|\mathbf{s}_1\|_2^2\} \le P_T, \mathbb{E}\{\|\hat{\mathbf{s}}^{(m)}\|_2^2\} \le P_T, \forall m, \\ &C_5 : p_{b_k} \ge 0,\ p_{u_k}^{(m)} \ge 0, \quad \forall k, m, \\ &C_6 : \mathbf{x}_{\min} \le \mathbf{x}_u^{(m)} \le \mathbf{x}_{\max}, \quad \forall m. \end{aligned} \tag{5.24}$$

This resulting problem in (5.24) is still non-convex due to the joint dependence of both the allocated powers p_{b_k}, $p_{u_k}^{(m)}$ and the UAV location $\mathbf{x}_u^{(m)} = [x_u^{(m)}, y_u^{(m)}]^T$ on the SINR expression in (5.6), which is used in the sum rate $\mathrm{R}_2^{(m)}$ calculation as given in (5.7). To overcome this challenge, we propose sequential optimization using swarm intelligence, which employs multiple agents, called particles, to explore the search space of objective function given in (5.24). Initially, Z_p particles are randomly placed in search space, where each particle communicates with other particles to share their personal best and update the current global best solution for the objective function. The particles then move iteratively for T iterations to reach the global optimum solution. In particular, each UAV location $\mathbf{x}_u^{(m)} = [x_u^{(m)}, y_u^{(m)}]^T$ and $\mathbf{P}_b, \mathbf{P}_u^{(m)}$ are optimized by using PSO-based algorithmic solution, while maximizing the total AR. Here, the zth particle at the tth iteration now represents an instance of each UAV location and MU PA matrices, which is given as follows:

$$\mathbf{J}_{p_z}^{(t)} = [\mathbf{X}_z^{(t)}, \hat{\mathbf{P}}_{b_z}^{(t)}, \hat{\mathbf{P}}_{u_z}^{(t)}]^T = [x_{1_z}^{(t)}, y_{1_z}^{(t)}, \cdots, x_{M_z}^{(t)}, y_{M_z}^{(t)}, \sqrt{\hat{p}_{b,1_z}^{(t)}}, \cdots, \sqrt{\hat{p}_{b,K_z}^{(t)}}, \cdots,$$
$$\cdots, \sqrt{\hat{p}_{u_1,1_z}^{(t)}}, \cdots, \sqrt{\hat{p}_{u_1,K_z}^{(t)}}, \cdots, \sqrt{\hat{p}_{u_M,1_z}^{(t)}}, \cdots, \sqrt{\hat{p}_{u_M,K_z}^{(t)}}]^T \in \mathbb{R}^{2K+2M}, \tag{5.25}$$

where each particle z represents the M_u UAV positions and PA to K IoT users from BS and each mth UAV, and calculates the objective function as $\mathrm{R}_T(\mathbf{F}_b, \mathbf{B}_b, \mathbf{F}_{u,t}^{(m)}, \mathbf{B}_{u,t}^{(m)}, \mathbf{F}_{u,r}^{(m)}, \mathbf{B}_{u,r}^{(m)}, \kappa_{b_z}^{(t)}\hat{\mathbf{P}}_{b_z}^{(t)}, \kappa_{u_z}^{(t)}\hat{\mathbf{P}}_{u_z}^{(t)}, \mathbf{X}_z^{(t)})$. We define $\hat{\mathbf{P}}_{b_z}^{(t)} = \mathrm{diag}(\sqrt{\hat{p}_{b,1_z}^{(t)}}, \cdots, \sqrt{\hat{p}_{b,K_z}^{(t)}}) \in \mathbb{R}^{K\times K}$ and $\hat{\mathbf{P}}_{u_z}^{(t)} = [\hat{\mathbf{P}}_{u_{1_z}}^{(t)}, \cdots, \hat{\mathbf{P}}_{u_{M_z}}^{(t)}]$, where $\hat{\mathbf{P}}_{u_{m_z}}^{(t)} = \mathrm{diag}(\sqrt{\hat{p}_{u_m,1_z}^{(t)}}, \cdots, \sqrt{\hat{p}_{u_m,K_z}^{(t)}}) \in \mathbb{R}^{K\times K}$ as the normalized PA matrices with $\hat{p}_{b,k_i}^{(t)}, \hat{p}_{u_m,k_z}^{(t)} \in [0, 1]$. Then, the transmit power constraints for $\hat{\mathbf{P}}_{b_z}^{(t)}$ and $\hat{\mathbf{P}}_{u_z}^{(t)}$ are satisfied by defining $\mathbf{P}_{b_z}^{(t)} = \kappa_{b_z}^{(t)}\hat{\mathbf{P}}_{b_z}^{(t)}$ and $\mathbf{P}_{u_z}^{(t)} = \kappa_{u_z}^{(t)}\hat{\mathbf{P}}_{u_z}^{(t)}$. The position $\mathbf{J}_{p_z}^{(t)}$ and velocity $\mathbf{J}_{v_z}^{(t)}$ for ith particle during tth iteration are updated as follows:

$$\mathbf{J}_{p_z}^{(t+1)} = \mathbf{J}_{p_z}^{(t)} + \mathbf{J}_{v_z}^{(t+1)}, \tag{5.26}$$

$$\mathbf{J}_{v_i}^{(t+1)} = \gamma_1 \mathbf{Y}_1^{(t)}(\mathbf{J}_{p_{\text{best}}}^{(t)} - \mathbf{J}_{p_i}^{(t)}) + \gamma_2 \mathbf{Y}_2^{(t)}(\mathbf{J}_{p_{\text{best}_i}}^{(t)} - \mathbf{J}_{p_i}^{(t)}) + \gamma_3^{(t)}\mathbf{J}_{v_i}^{(t)}, \tag{5.27}$$

where γ_1 and γ_2 are the learning parameters for the global best $\mathbf{J}_{p_{\text{best}}}^{(t)}$ and the personal best $\mathbf{J}_{p_{\text{best}_i}}^{(t)}$, $\gamma_3^{(t)} = \Gamma_l - (t/T)(\Gamma_u - \Gamma_l)$ is the inertia parameters with the upper bound Γ_u and lower bound Γ_l for decreasing the velocity as the number of iterations increases, and $\mathbf{Y}_1^{(t)}$ and $\mathbf{Y}_2^{(t)}$ are the random diagonal matrices with the uniformly distributed entries over $[0, 1]$. Finally, the personal and global best solutions for zth particle during tth iteration are obtained as follows:

$$\mathbf{J}_{p_{\text{best}_z}}^{(t)} = \underset{\mathbf{J}_{p_z}^{(t^*)}, \forall t^*=0,1,\cdots,t}{\arg\max} \; R_T(\mathbf{F}_b, \mathbf{B}_b, \mathbf{F}_{u,t}^{(m)}, \mathbf{B}_{u,t}^{(m)}, \mathbf{F}_{u,r}^{(m)}, \cdots,$$
$$\cdots, \mathbf{B}_{u,r}^{(m)}, \kappa_{b_z}^{(t^*)}\hat{\mathbf{P}}_{b_z}^{(t^*)}, \kappa_{u_z}^{(t^*)}\hat{\mathbf{P}}_{u_z}^{(t^*)}, \mathbf{X}_z^{(t^*)}), \tag{5.28}$$

$$\mathbf{J}_{p_{\text{best}}}^{(t)} = \underset{\mathbf{J}_{p_{\text{best}_z}}^{(t)}, \forall z=0,1,\cdots,Z_p}{\arg\max} \; R_T(\mathbf{F}_b, \mathbf{B}_b, \mathbf{F}_{u,t}^{(m)}, \mathbf{B}_{u,t}^{(m)}, \mathbf{F}_{u,r}^{(m)}, \cdots,$$
$$\cdots, \mathbf{B}_{u,r}^{(m)}, \kappa_{\text{best},b_z}^{(t)}\hat{\mathbf{P}}_{\text{best},\text{b}_z}^{(t)}, \kappa_{\text{best},u_z}^{(t)}\hat{\mathbf{P}}_{\text{best},\text{u}_z}^{(t)}, \mathbf{X}_z^{(t)}). \tag{5.29}$$

After T iterations, we update $\mathbf{x}_u^{(m)} = \mathbf{X}_{\text{best}}^{(T)}$, $\mathbf{P}_b = \kappa_{\text{best},b}^{(T)}\hat{\mathbf{P}}_{\text{best},b}^{(T)}$, and $\mathbf{P}_u^{(m)} = \kappa_{\text{best},u}^{(T)}\hat{\mathbf{P}}_{\text{best},u}^{(T)}$.

Algorithm 5.1: Proposed joint HBF, UAV location optimization, and PA for multiple UAVs (J-HBF-PSOPA-FL) algorithm

Input: Z_p, T, (θ, ψ), (x_b, y_b, z_b), $(x_u^{(m)}, y_u^{(m)}, z_u^{(m)})$.
Output: $\mathbf{x}_u^{(m)}$, $\mathbf{P}_b$, $\mathbf{P}_u$, $\mathbf{F}_b$, $\mathbf{B}_b$, $\mathbf{F}_{u,t}$, $\mathbf{B}_{u,t}$, $\mathbf{F}_{u,r}$, $\mathbf{B}_{u,r}$.
Formulate BS RF and BB stages using (5.14), (5.23).
Formulate UAVs receive HBF stages using (5.14), (5.20).
Construct UAV transmit HBF stages via (5.14), (5.21).
for $i = 1 : Z_p$ **do**
    Initialize the velocity as $\mathbf{J}_{v_i}^{(0)} = \mathbf{0}$.
    Each entry of $\mathbf{J}_{p_i}^{(0)}$ is uniformly distributed in [0, 1].
    Set the personal best $\mathbf{J}_{p\text{best}_i}^{(0)} = \mathbf{J}_{p_i}^{(0)}$.
end
Find the global best $\mathbf{J}_{p\text{best}}^{(0)}$ as in (4.40).
for $t = 1 : T$ **do**
    for $i = 1 : Z_p$ **do**
        Update the velocity $\mathbf{J}_{v_i}^{(t)}$ as in (5.27).
        Update the position $\mathbf{J}_{p_i}^{(t)}$ as in (5.26).
        Find the personal best $\mathbf{J}_{p\text{best}_i}^{(t)}$ as in (5.28).
    end
    Find the global best $\mathbf{J}_{p\text{best}}^{(t)}$ as in (5.29).
end
$\mathbf{x}_u^{(m)} = \mathbf{X}_{\text{best}}^{(T)}$, $\mathbf{P}_b = \kappa_{\text{best},b}^{(T)} \hat{\mathbf{P}}_{\text{best},b}^{(T)}$, $\mathbf{P}_u^{(m)} = \kappa_{\text{best},u}^{(T)} \hat{\mathbf{P}}_{\text{best},u}^{(T)}$.

5.4 Illustrative Results

In this section, the Monte Carlo simulation results are presented based on the proposed scheme. Table 5.1 outlines the simulation setup based on the 3-D microcell scenario [10] for the results discussed hereafter. The PSO parameters are chosen as: $Z_p = 20$, $\gamma_1 = \gamma_2 = 2$, and $\gamma_3 = 1.1$. In Fig. 5.3a, we compare the AR R_2 versus transmit power for the following two cases: (1) a single UAV ($M = 1$) deployed at initial fixed location $(x_u, y_u) = (50, 50)$ and (2) multiple UAVs ($M = 2$) deployed at initial fixed locations $(x_u^{(1)}, y_u^{(1)}) = (50, 50)$, $(x_u^{(2)}, y_u^{(2)}) = (100, 50)$. We consider a practical user distribution scenario where the users are located at multiple locations (i.e., $(x_k, y_k) \in [50,100]$) from BS, which is located at $(x_b, y_b) = (0, 0)$. It can be seen from Fig. 5.3a that the proposed scheme can optimize the UAV location; however, it can only achieve a suboptimal solution as a single UAV cannot be positioned optimally to support a large number of users. Moreover, each kth user experiences interference from $K - 1$ users, which leads to a low AR. To improve the performance, a multiple UAV-assisted system (i.e., $M = 2$) is used in Fig. 5.3b, where each UAV can support $K_m = K/M$ users based on the proposed scheme and find the optimal deployments close to its associated K_m users. It can be seen that using the proposed J-HBF-PSOLPA scheme, each UAV can optimally cluster its users and then find the optimal deployment while achieving the maximum AR.

Table 5.1 Simulation parameters

Number of antennas		$(N_T, N_r, N_t) = 64$	
Number of paths	Path loss exponent	$L = 10$	3.6
Frequency	Channel bandwidth	28 GHz	100 MHz
Noise PSD	Number of UAVs M_u	− 174 dBm/Hz	2 or 3
BS	UAV height	10 m	20 m
UAV x-axis range	UAV y-axis range	$[x_{\min}, x_{\max}] =$ [0, 100] m	$[y_{\min}, y_{\max}] =$ [0, 100] m
Azimuth AoD/AoA (1st link)	Azimuth AoD/AoA (2nd link)	120°	150°
Elevation AoD/AoA (1st link)	Elevation AoD/AoA (2nd link)	60°	30°
Azimuth/elevation angle spread	# of network realizations	± 10°	2000

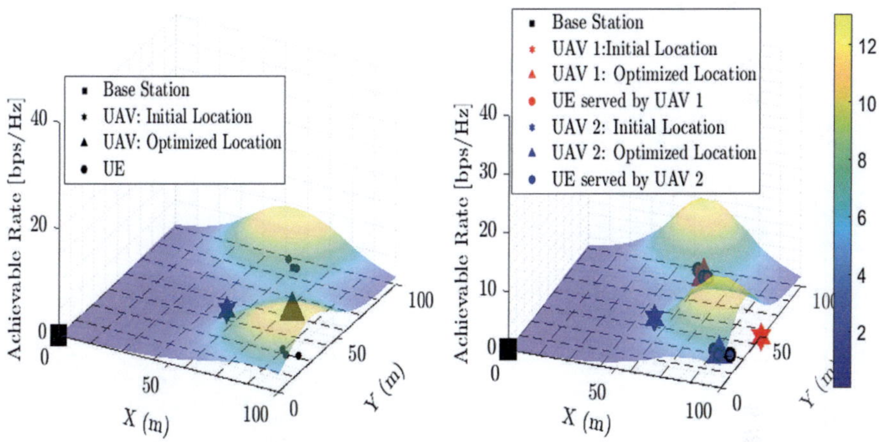

(a) (b)

Fig. 5.3 AR R_2 versus $(x - y)$ coordinates at $P_T^{(m)} = 20$ dBm. (**a**) Single UAV deployment ($M_u = 1$). (**b**) Multiple UAV deployment ($M_u = 2$)

Thus, a multiple UAV-assisted MU-mMIMO system can overcome the coverage and capacity limitation of a single UAV in more practical scenarios.

Figure 5.4 compares the AR of a multiple UAV system versus a single UAV system for the following cases: (1) J-HBF-PSOLPA with direct link from BS (we consider optimal PA at each UAV, while considering equal PA from BS); (2) J-HBF-PSOLPA without direct link from BS; and (3) J-HBF-FL-EQPA. It can be seen that by using $M = 2$ UAV can provide approximately 2.5 times the AR when compared to a single UAV case ($M = 1$). For instance, the AR at $P_T = 50$ dBm is increased from 27 bps/Hz to 58 bps/Hz, which indicates around 250% increase in total AR. Moreover, compared to J-HBF-FL-EQPA, the proposed scheme can significantly enhance the performance by optimizing the UAV locations and the

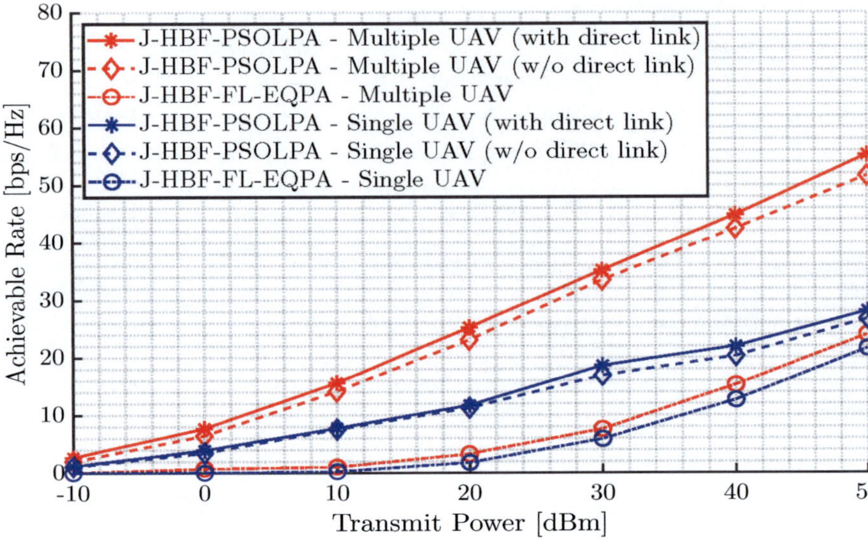

Fig. 5.4 Total AR R_T versus P_T for single and multiple UAV system

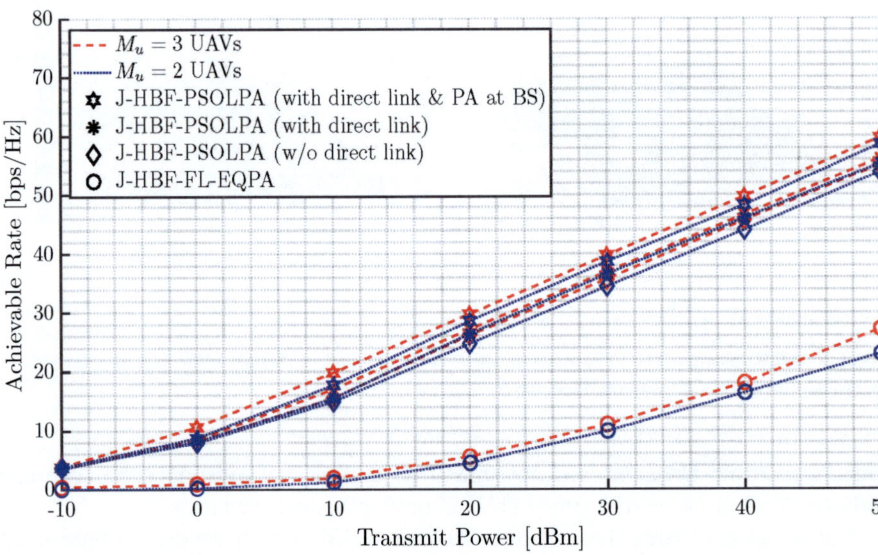

Fig. 5.5 Total AR R_T versus P_T for 2 and 3 UAV systems

allocated powers. The inclusion of a direct link from BS to users (i.e., cooperative relaying) can provide an additional 4 to 5 bps/Hz rate improvement for all cases. Figure 5.5 compares the AR for different numbers of UAVs (i.e., for $M = 2$ or $M = 3$) for the following four cases: (1) J-HBF-PSOLPA with direct link and optimal PA at BS and each UAV; and (2) J-HBF-PSOLPA with direct link and optimal PA at each UAV only; (3) J-HBF-PSOLPA without direct link; and (4) J-HBF-FL-EQPA. The analysis can be summarized as follows: (1) the optimal PA from BS and each UAV can provide an improved performance (e.g., an increase of $\approx 10\%$ rate) when compared to only optimal PA at each UAV; (2) by increasing the number of UAVs, we can achieve a higher AR (e.g., the rate can be increased by 5–7% when M_u is increased from 2 to 3). Moreover, by increasing the number of UAVs, we can further improve the performance as each UAV cluster has fewer users, which leads to reduced MU interference and thus, an increased AR. However, it must be noted that optimizing the number of UAVs is beyond the scope of this research and is left as our future work.

5.5 Concluding Remarks

In this chapter, we considered a MU-mMIMO IoT cooperative relaying system, where multiple UAV DF relays connect the BS to a large number of users. We have proposed a sequential optimization scheme that employs swarm intelligence to assign users to UAVs through K-means clustering and optimizes the UAV locations and PA from the BS and each UAV, followed by the design of RF and BB stages for maximum achievable rates. The RF stages are designed using the angular information of UAVs and users, while the BB stages are designed using reduced-dimension effective channel matrices. Our findings show that multiple UAV-assisted cooperative relaying system works better than a single UAV, especially when taking into account the practical user distributions. Moreover, compared to fixed positions and equal power distribution of UAV, the optimization of UAV locations and PA substantially improves the AR.

References

1. Y. Chen, N. Zhao, Z. Ding, M.-S. Alouini, Multiple UAVs as relays: multi-hop single link versus multiple dual-hop links. IEEE Trans. Wirel. Commun. **17**(9), 6348–6359 (2018)
2. C. Zhang, L. Zhang, L. Zhu, T. Zhang, Z. Xiao, X.-G. Xia, 3D deployment of multiple UAV-mounted base stations for UAV communications. IEEE Trans. Commun. **69**(4), 2473–2488 (2021)
3. N. Nouri, F. Fazel, J. Abouei, K.N. Plataniotis, Multi-UAV placement and user association in uplink MIMO ultra-dense wireless networks. IEEE Trans. Mobile Comput. **22**(3), 1615–1632 (2023)
4. M. Mahmood, A. Koc, T. Le-Ngoc, PSO-based joint UAV positioning and hybrid precoding in UAV-assisted massive MIMO systems, in *Proceedings of the IEEE 96th Vehicular Technology Conference (VTC-Fall)* (2022), pp. 1–6
5. J. Du, W. Xu, Y. Deng, A. Nallanathan, L. Vandendorpe, Energy-saving UAV-assisted multiuser communications with massive MIMO hybrid beamforming. IEEE Commun. Lett. **24**(5), 1100–1104 (2020)

6. M. Mahmood, A. Koc, T. Le-Ngoc, Spherical array-based joint beamforming and UAV positioning in massive MIMO systems, in *Proceedings of the IEEE 97th Vehicular Technology Conference (VTC-Spring)* (2023), pp. 1–5
7. X. Jiang, Z. Wu, Z. Yin, Z. Yang, Power and trajectory optimization for UAV-enabled Amplify-and-Forward relay networks. IEEE Access **6**, 48688–48696 (2018)
8. K. Xiong, P. Fan, Z. Xu, H.-C. Yang, K.B. Letaief, Optimal cooperative beamforming design for MIMO decode-and-forward relay channels. IEEE Trans. Signal Process. **62**(6), 1476–1489 (2014)
9. R. Méndez-Rial, C. Rusu, N. Gonzàlez-Prelcic, A. Alkhateeb, R. W. Heath, Hybrid MIMO architectures for millimeter wave communications: phase shifters or switches? IEEE Access **4**, 247–267 (2016)
10. M. Mahmood, M. Ghadaksaz, A. Koc, T. Le-Ngoc, Deep learning meets swarm intelligence for UAV-assisted IoT coverage in massive MIMO. IEEE Internet Things J. **11**(5), 7679–7696 (2024)

Part II
Low-Complexity DL-Based Solutions for Real-Time UAV-Assisted MU-mMIMO Communications

Chapter 6
Joint UAV Location and PA Optimization: A DSL Approach

6.1 Introduction

In Part I of this book, we explored various PSO-based algorithmic solutions combined with HBF to reduce the MU interference for different UAV-assisted mMIMO communications problems to maximize AR, where it was seen that PSO-based solutions achieved near-optimal AR in different scenarios. However, while these PSO-based solutions accomplish an acceptable performance, they have high computational complexity and struggle in real-time scenarios. Therefore, in Part II of this book, we aim to investigate low-complexity DL-based solutions to reduce the runtime of the proposed algorithms without significantly affecting the performance.

In Sect. 2.5.2, we provided the background and related literature in UAV-assisted communications for using SL [1–4] and DSL [5–7] to reduce the computational complexity. Motivated by this, in this chapter, we aim to reduce the complexity of the proposed J-HBF-PSOLPA algorithm in Chap. 4. To overcome the high computational complexity of J-HBF-PSOLPA, we propose a novel low-complexity J-HBF-DSLLPA algorithm for a UAV-assisted MU-mMIMO communications systems. The proposed J-HBF-DSLLPA-based solution can achieve AR of J-HBF-PSOLPA, while reducing the runtime by 98–99%. In particular, the proposed J-HBF-DSLLPA is built via a fully connected DNN consisting of two phases: (1) offline SL via the optimal allocated powers and UAV locations calculated with J-HBF-PSOLPA; and (2) real-time prediction of optimal power values and UAV location via the trained DNN.

The rest of this chapter is organized as follows. Section 6.2 presents the system and channel model of UAV-assisted MU-mMIMO systems followed by the proposed low-complexity DSL-based joint HBF, PA, and UAV location optimization scheme in Sect. 6.3. The illustrative results are in Sect. 6.4 followed by the conclusions in Sect. 6.5.

T. Le-Ngoc et al., *AI-Enabled UAV-Assisted Massive MIMO*, Wireless Networks,
https://doi.org/10.1007/978-3-032-07872-8_6

6.2 System and Channel Model

We consider the same system model introduced in Chap. 4, where the UAV operates as a DF relay between BS and several remote users/devices. Similarly, the only means of communication between the users and the BS is the UAV relay. In particular, The BS has N_T antennas for transmission, the UAV has N_r for reception, and N_t for transmission to K single antenna users clustered in G groups. Each group g has K_g users such that $K = \sum_{g=1}^{G} K_g$. Considering half-duplex communication HBF and for both BS and UAV, the BS sends the $N_s = K$ information data $\mathbf{d} = [d_1, \cdots, d_K]^T$ that is transmitted using the BS's BB stage $\mathbf{B}_b \in \mathbb{C}^{N_{RF_b} \times K}$ and RF stage $\mathbf{F}_b \in \mathbb{C}^{N_T \times N_{RF_b}}$ to the UAV and received by the UAV's receive RF stage $\mathbf{F}_{u,r} \in \mathbb{C}^{N_r \times N_T}$ and BB stage $\mathbf{B}_{u,r} \in \mathbb{C}^{K \times N_{RF_u}}$ in the first time slot. Then, the UAV transmits K data streams using UAV's transmit BB stage $\mathbf{B}_{u,t} \in \mathbb{C}^{N_{RF_u} \times K}$ and MU PA matrix $\mathbf{P} = \text{diag}\left(\sqrt{p_1}, \cdots, \sqrt{p_K}\right) \in \mathbb{C}^{K \times K}$ and UAV's transmit RF stage $\mathbf{F}_{u,t} \in \mathbb{C}^{N_t \times N_{RF_u}}$ to K users in the second time slot. Here, N_{RF_b} and N_{RF_u} are the number of RF chains at the BS and UAV, respectively, such that $K \leq N_{RF_b} \leq N_T$ and $K \leq N_{RF_u} \leq N_r$ (N_t). Subsequently, as explained in Sect. 4.2.1, the AR for the link between the BS and UAV R_1 can be written as

$$\mathrm{R}_1(\mathbf{F}_b, \mathbf{B}_b, \mathbf{F}_{u,r}, \mathbf{B}_{u,r}) = \log_2 |\mathbf{I}_K + \mathbf{Q}_1^{-1}\mathbf{B}_{u,r}\mathcal{H}_1\mathbf{B}_b\mathbf{B}_b^H\mathcal{H}_1^H\mathbf{B}_{u,r}^H|. \tag{6.1}$$

Here, $\mathbf{I}_K = \mathbb{E}\left\{\mathbf{d}\mathbf{d}^H\right\} \in \mathbb{C}^{K \times K}$, $\mathbf{Q}_1^{-1} = (\sigma_n^2 \mathbf{B}_{u,r}\mathbf{F}_{u,r})^{-1}\mathbf{F}_{u,r}^H\mathbf{B}_{u,r}^H$, $\mathcal{H}_1 = \mathbf{F}_{u,r}\mathbf{H}_1\mathbf{F}_b$, and σ_n^2 is the noise power. Subsequently, the SINR_{g_k}, the SINR for the kth user in the gth group, is expressed as

$$\text{SINR}_{g_k} = \frac{p_{g_k} |\mathbf{h}_{2,k}^H \mathbf{F}_{u,t} \mathbf{b}_{u,t,g_k}|^2}{\sum\limits_{\hat{k} \neq k}^{K_g} p_{g_{\hat{k}}} |\mathbf{h}_{2,k}^H \mathbf{F}_{u,t} \mathbf{b}_{u,t,g_{\hat{k}}}|^2 + \sum\limits_{q \neq g}^{G} \sum\limits_{\hat{k} \neq k}^{K_g} p_{q_{\hat{k}}} |\mathbf{h}_{2,k}^H \mathbf{F}_{u,t} \mathbf{b}_{u,t,q_{\hat{k}}}|^2 + \sigma_n^2}, \tag{6.2}$$

which is used to calculate the sum rate for the link between UAV and users R_2 as

$$\mathrm{R}_2(\mathbf{F}_{u,t}, \mathbf{B}_{u,t}, \mathbf{P}, x_u, y_u) = \mathbb{E}\left\{\sum_{g=1}^{G}\sum_{k=1}^{K_g} \mathbb{E}\left[\log_2(1 + \text{SINR}_{g_k})\right]\right\}. \tag{6.3}$$

In equations above, $\mathbf{H}_1 \in \mathbb{C}^{N_r \times N_T}$, $\mathbf{H}_2 \in \mathbb{C}^{K \times N_r}$, and $\mathbf{h}_{2,k} \in \mathbb{C}^{N_t}$ represent the channel between BS and UAV, between UAV and users, and between UAV and kth users, respectively. The channel models are discussed in detail in Sect. 4.2.2.

6.3 DSL-Based Joint UAV Deployment and PA

We investigate the problem of joint HBF, UAV location, and PA optimization problem stated in (4.14). After the formulation of RF beamformers $\mathbf{F}_b$, $\mathbf{F}_{u,r}$, $\mathbf{F}_{u,t}$ (formulated in Sect. 4.3.1), and BB stages $\mathbf{B}_b$, $\mathbf{B}_{u,t}$, $\mathbf{B}_{u,r}$ (formulated in Sect. 4.3.2), the optimization problem for maximum AR can be formulated as follows:

$$\begin{aligned}
&\max_{\{\mathbf{P},\mathbf{x}_u\}} \quad \mathrm{R_T}\left(\mathbf{F}_b, \mathbf{B}_b, \mathbf{F}_{u,r}, \mathbf{B}_{u,r}, \mathbf{F}_{u,t}, \mathbf{B}_{u,t}, \mathbf{P}, \mathbf{x}_u\right) \\
&\text{s.t.} \quad C_4 : \mathbb{E}\left\{\|\mathbf{s}_2\|_2^2\right\} \leq \sum_{k=1}^{K} p_k \mathbf{b}_{u,t,k}^H \mathbf{F}_{u,t}^H \mathbf{F}_{u,t} \mathbf{F}_{u,t} \mathbf{b}_{u,t,k}, \leq P_T \\
&\qquad C_5 : p_k \geq 0 \quad \forall k \\
&\qquad C_6 : x_{\min} \leq x_u \leq x_{\max}, \, y_{\min} \leq y_u \leq y_{\max},
\end{aligned} \tag{6.4}$$

where $\mathrm{R}_T = \frac{1}{2}\min(\mathrm{R}_1, \mathrm{R}_2)$. The optimization problem in (6.4) was solved in Sect. 4.3.3.3 using a PSO-based algorithm namely J-HBF-PSOLPA, where it was shown that it can achieve a solution close to optimal. However, as the number of users increases (i.e., higher dimensional optimization space), the proposed PSO-based solution requires more iterations and longer runtime. Thus, the enhanced computational complexity might make PSO-based solutions impractical for the real-time online applications of MU-mMIMO systems.

To address this challenge, we propose a low-complexity DSL-based algorithm, called J-HBF-DSLLPA, which can achieve a near-optimal AR, while maintaining a reasonable runtime. The proposed algorithm has two phases, as illustrated in Fig. 6.1: (1) Phase 1 applies the offline SL via the optimal allocated power and

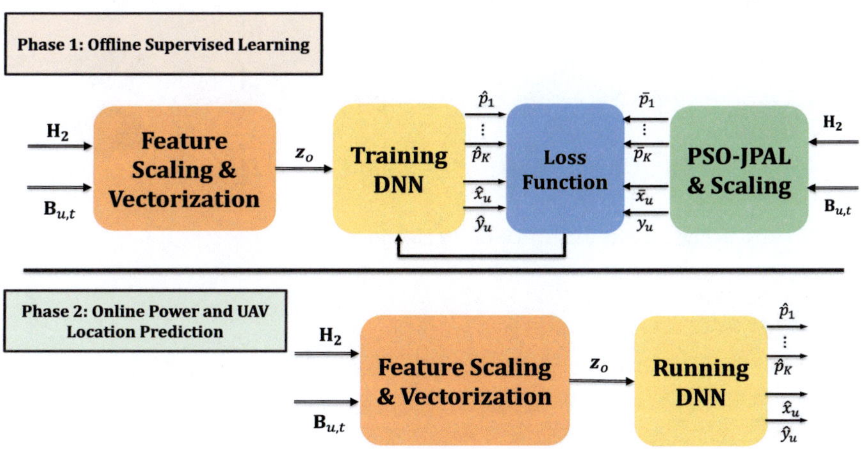

Fig. 6.1 Block diagram of offline SL and real-time prediction in J-HBF-DSLLPA algorithm

location values calculated by J-HBF-PSOLPA, and (2) Phase 2 runs the trained J-HBF-DSLLPA algorithm for predicting the allocated powers and location in the real-time online applications. Therefore, the remaining part of this section focuses on the DNN architecture, loss functions, dataset preparation, and training process for the proposed low-complexity J-HBF-DSLLPA algorithm.

6.3.1 Proposed DNN Architecture

We employ a fully connected DNN architecture with four HLs as depicted in Fig. 6.2, which aims to predict the optimal allocated powers for K users as well as optimal location in a given deployment span. We consider ζ_l neurons in each lth HL with $l = 1, \cdots, 4$. As shown in Fig. 6.2, the proposed J-HBF-DSLLPA algorithm uses the effective channel matrix between K users $\mathcal{H}_2 \in \mathbb{C}^{K \times N_{RF_u}}$ given in (4.24), and the transmit BB stage $\mathbf{B}_{u,t} \in \mathbb{C}^{N_{RF_u} \times K}$ given in (4.33) as inputs, which are first subject to feature scaling and vectorization operations. The choice of $\mathcal{H}_2$ and $\mathbf{B}_{u,t}$ as inputs of the network ensures that sufficient information regarding the user configuration is given to the network to predict the location and PA without

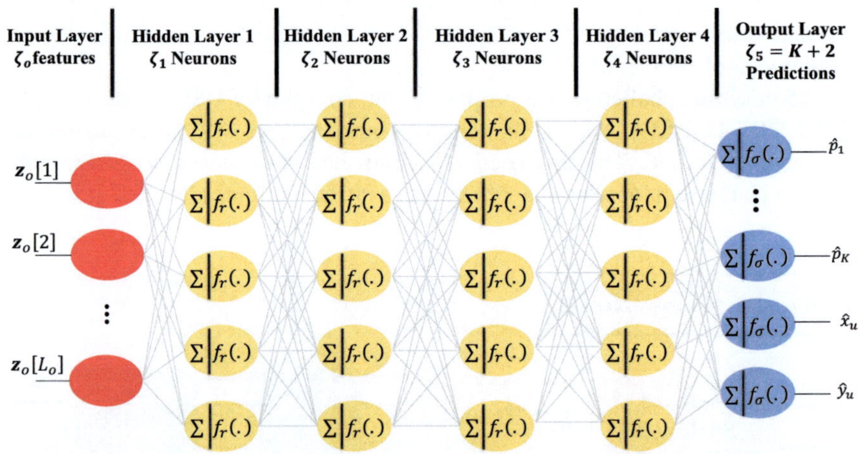

Fig. 6.2 DNN architecture for J-HBF-DSLLPA algorithm

explicitly giving the users' locations. Subsequently, the input layer feature vector is derived as

$$\mathbf{z}_0 = \begin{bmatrix} \omega_1 \mathbf{z}_{\tilde{\mathbf{h}}_{2,1}} \\ \vdots \\ \omega_1 \mathbf{z}_{\tilde{\mathbf{h}}_{2,K}} \\ \omega_2 \mathbf{z}_{\mathbf{b}_{u,t,1}} \\ \vdots \\ \omega_2 \mathbf{z}_{\mathbf{b}_{u,t,K}} \\ \omega_3 \mathbf{z}_{\mathbf{BB}_{u,t}} \\ \omega_4 \mathbf{z}_{\mathbf{BB,inv,u,t}} \end{bmatrix} \in \mathbb{R}^{\zeta_0}, \tag{6.5}$$

where $\zeta_0 = (2N_t + 2N_{RF_u} + 2)K$ is the input feature size, $\mathbf{z}_{\tilde{\mathbf{h}}_{2,k}} = [\mathrm{Re}(\tilde{\mathbf{h}}_{2,k}^T), \mathrm{Im}(\tilde{\mathbf{h}}_{2,k}^T)]^T \in \mathbb{R}^{2N_t}$, $\mathbf{z}_{\mathbf{b}_{u,t,k}} = [\mathrm{Re}(\mathbf{b}_{u,t,k}^T), \mathrm{Im}(\mathbf{b}_{u,t,k}^T)]^T \in \mathbb{R}^{2N_{RF_u}}$, $\mathbf{z}_{\mathbf{BB}_{u,t}^T} = [\mathbf{b}_{u,t,1}^H \mathbf{b}_{u,t,1}, \cdots, \mathbf{b}_{u,t,K}^H \mathbf{b}_{u,t,K}]^T \in \mathbb{R}^K$, and $\mathbf{x}_{\mathbf{BB,inv,u,t}^T} = [\frac{1}{\mathbf{b}_{u,t,1}^H \mathbf{b}_{u,t,1}}, \cdots, \frac{1}{\mathbf{b}_{u,t,K}^H \mathbf{b}_{u,t,K}}]^T \in \mathbb{R}^K$ are respectively the non-scaled input feature vectors for the effective channel, transmit BB stage, the gain of each BB precoder vector, and its inverse. By implementing the maximum absolute scaling [8], the corresponding scaling coefficients are calculated as

$$\begin{aligned} \omega_1 &= \max\left(|\mathbf{z}_{\tilde{\mathbf{h}}_{2,1}}^T|, \cdots, |\mathbf{z}_{\tilde{\mathbf{h}}_{2,K}}^T|\right)^{-1} \\ \omega_2 &= \max\left(|\mathbf{z}_{\tilde{\mathbf{b}}_{u,t,1}}^T|, \cdots, |\mathbf{z}_{\tilde{\mathbf{b}}_{u,t,K}}^T|\right)^{-1} \\ \omega_3 &= \max\left(\mathbf{b}_{u,t,1}^H \mathbf{b}_{u,t,1}^H, \cdots, \mathbf{b}_{u,t,K}^H \mathbf{b}_{u,t,K}^H\right)^{-1} \\ \omega_4 &= \max\left(\mathbf{b}_{u,t,1}^H \mathbf{b}_{u,t,1}^H, \cdots, \mathbf{b}_{u,t,K}^H \mathbf{b}_{u,t,K}^H\right). \end{aligned} \tag{6.6}$$

The proposed algorithm utilizes the maximum absolute scaling technique to scale the input feature vector between -1 and 1 (i.e., $\mathbf{z}_0 \in (-1, 1]$), which prevents certain features from dominating the learning process. In the offline SL process, the optimal PA and UAV location values are calculated as the output labels via the computationally expensive J-HBF-PSOLPA algorithm. Similar to the input features, we also apply the maximum absolute scaling to the optimal allocated powers and UAV location as follows:

$$\bar{x}_u = \frac{x_u^{\mathrm{opt}}}{\max(x_u, y_u)} \in [0, 1] \tag{6.7}$$

$$\bar{y}_u = \frac{y_u^{\mathrm{opt}}}{\max(x_u, y_u)} \in [0, 1] \tag{6.8}$$

$$\bar{p}_k = \frac{p_k^{\text{opt}}}{\max(p_1^{\text{opt}}, \cdots, p_K^{\text{opt}})} \in [0, 1]. \tag{6.9}$$

To perform nonlinear operations, we adopt the ReLU as the activation function in the HLs (i.e., $f_r(z) = \max(0, z)$). Therefore, using the input feature vector of $\mathbf{z}_0$ given in (6.5), the output of lth HL is computed as $\mathbf{z}_l = f_r(\mathbf{W}_{l-1}\mathbf{z}_{l-1} + \mathbf{b}_{l-1}) \in \mathbb{R}^{\zeta_l}$, where $\mathbf{W}_{l-1} \in \mathbb{R}^{\zeta_l \times \zeta_{l-1}}$ and $\mathbf{b}_{l-1} \in \mathbb{R}^{\zeta_l}$ are the weight matrix and bias vector, respectively. To ensure that the predicted output values are between 0 and 1, we apply the sigmoid function at the output layer (i.e., $f_\sigma(z) = \frac{1}{1+e^{-z}}$). Thus, the predicted power and UAV location values via the DNN architecture are written as follows:

$$\begin{aligned}
&\left[\hat{p}_1, \hat{p}_2, \cdots, \hat{p}_K, \hat{x}_o, \hat{y}_o\right] \\
&= f_\sigma(\mathbf{W}_4\mathbf{z}_4 + \mathbf{b}_4) \\
&= f_\sigma(\mathbf{W}_4 f_r(\mathbf{W}_3 f_r(\mathbf{W}_2 f_r(\mathbf{W}_1 f_r(\mathbf{W}_0\mathbf{x}_0 + \mathbf{b}_0) + \mathbf{b}_1) + \mathbf{b}_2) + \mathbf{b}_3) + \mathbf{b}_4).
\end{aligned} \tag{6.10}$$

By using (4.33) and (6.10), we derive the UAV PA matrix satisfying the transmit power constraint of P_T (i.e., C_4 in (6.4)) as

$$\mathbf{P} = \sqrt{\frac{P_T}{\sum_{k=1}^{K} \hat{p}_k \mathbf{b}_k^H \mathbf{b}_k}} \operatorname{diag}\left(\hat{p}_1, \cdots, \hat{p}_K\right). \tag{6.11}$$

6.3.2 Loss Function

We here consider two loss functions by using the predicted and optimal power values: (1) MSE and (2) MAE. Considering a minibatch gradient descent approach [9] with a batch size = B_S, the MSE loss function is given by

$$\mathcal{L}_{\text{MSE}} = \frac{1}{B_S K} \sum_{i=1}^{B_S} \sum_{k=1}^{K} (\bar{p}_{k,i} - \hat{p}_{k,i})^2 + \frac{1}{B_S} \sum_{i=1}^{B_S} (\bar{x}_u - \hat{x}_u)^2 + \frac{1}{B_S} \sum_{i=1}^{B_S} (\bar{y}_u - \hat{y}_u)^2. \tag{6.12}$$

Similarly, the MAE loss function is written as

$$\mathcal{L}_{\text{MAE}} = \frac{1}{B_S K} \sum_{i=1}^{B_S} \sum_{k=1}^{K} |\bar{p}_{k,i} - \hat{p}_{k,i}| + \frac{1}{B_S} \sum_{i=1}^{B_S} |\bar{x}_u - \hat{x}_u| + \frac{1}{B_S} \sum_{i=1}^{B_S} |\bar{y}_u - \hat{y}_u|. \tag{6.13}$$

The weight matrices $\mathbf{W}_l$ and bias vectors $\mathbf{b}_l$ of the DNN architecture are updated by back-propagating the gradients of the loss function from the output layer to the input layer. This helps in reducing the loss and accurately predicting the optimal allocated power and UAV location values, thus maximizing the total AR as expressed in (6.4).

6.3.3 *Data Generation and Training Process*

We generated a dataset of $S_r = 100.000 = 10^5$ network realizations to train the proposed DNN architecture, as shown in Fig. 6.2. For each realization, we randomly varied the path gains, AoD parameters, and UE location to generate the channel vector for each UE as given in (4.12). The corresponding optimal allocated powers and UAV location are calculated via the J-HBF-PSOLPA algorithm and stored in the dataset. The total available dataset was split into 80% for training and 20% for validation for the offline SL process (i.e., phase 1). After completing the offline SL process, the online PA and UAV location (i.e., phase 2) are tested with a purely new test dataset. The DNN architecture for the proposed J-HBF-DSLLPA algorithm is implemented using the open-source DL libraries in TensorFlow [10].

6.4 Illustrative Results

In this section, we present the Monte Carlo simulation results based on the proposed algorithmic solutions. Table 6.1 outlines the simulation setup based on the 3-D microcell scenario [11] for the results discussed hereafter.

Table 6.1 Simulation parameters

Number of antennas		$(N_T, N_t, N_r) = 144$	
Number of paths	Path loss exponent	$L = 10$	3.6
Frequency	Channel bandwidth	28 GHz	100 MHz
Noise PSD	Reference path loss α	− 174 dBm/Hz	61.34 dB
BS/Gateway height	UAV height	10 m	20 m
User groups	# of users per group	$G = 1$ or 2	$K_g = \frac{K}{G}$
UAV x-axis range	UAV y-axis range	$[x_{\min}, x_{\max}] =$ [0, 100] m	$[y_{\min}, y_{\max}] =$ [0, 100] m
Mean AAoD/AAoA (1st link)	Mean AAoD (2nd link)	120°	$\phi_g = 21° + 120°(g-1)$
Mean EAoD/EAoA (1st link)	Mean EAoD (2nd link)	60°	$\theta_g = 60°$
Azimuth/elevation angle spread	# of network realizations	±10°	2000
Minimum horizontal distance $\tau_{h,\min}$	Maximum horizontal distance $\tau_{h,\min}$	0	100 [m]

Table 6.2 DNN parameters

Regularizer		L2	
1st hidden layer size	2nd hidden layer size	$\zeta_1 = 1024$	$\zeta_2 = 512$
3rd hidden layer size	4th hidden layer size	$\zeta_3 = 256$	$\zeta_4 = 128$
Optimizer	Learning rate	ADAM	0.001
Dataset size	Test data	$S_r = 100{,}000$	$S_t = 1000$
Batch size	Epoch size	32	15

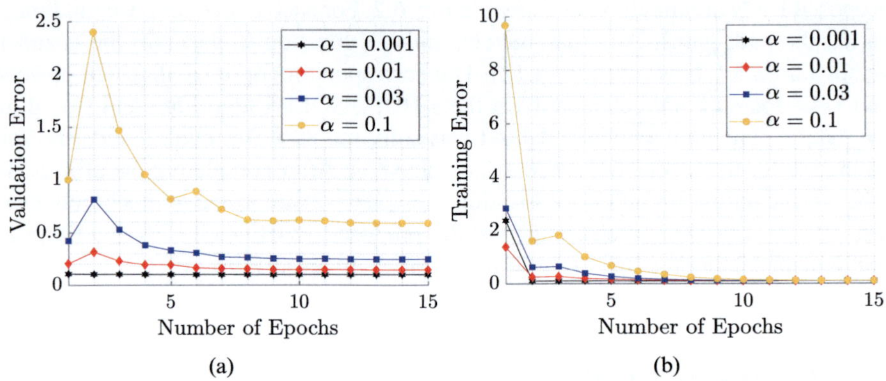

Fig. 6.3 MSE loss vs. number of epochs for different learning rates: (**a**) Validation error. (**b**) Training error

6.4.1 *Low-Complexity DSL-Based Performance*

In this section, we present the results obtained using the proposed low-complexity J-HBF-DSLLPA algorithm and compare its performance with that of J-HBF-PSOLPA. Table 6.2 outlines the hyper-parameters for the proposed DNN architecture, which is given in Fig. 6.2. Figure 6.3 shows the MSE performance for training and validation datasets using different learning rates $\alpha = \{0.001, 0.01, 0.03, 0.1\}$ over a total of 15 epochs. Higher learning rates, such as 0.1 and 0.03, lead to larger validation errors and take more iterations to stabilize. On the other hand, smaller learning rates like $\alpha = 0.01$ and 0.001 result in lower validation errors, even with fewer epochs. Specifically, the performance follows the pattern $\text{MSE}_{\alpha=0.001} < \text{MSE}_{\alpha=0.01} < \text{MSE}_{\alpha=0.03} \ll \text{MSE}_{\alpha=0.1}$, highlighting the need to select an appropriate learning rate. Similarly, Fig. 6.4 shows the MSE performance for training and validation datasets using a different number of HLs in DNN architecture. The error is evaluated over 15 epochs for $\text{HL} = \{2, 3, 4\}$. The results show that DNNs with 2 or 3 HLs can achieve low MSE, but they may take more epochs to reach convergence. Using 4 HLs, however, provides better MSE performance for the validation dataset, following the trend ($\text{MSE}_{\text{HL}=4} < \text{MSE}_{\text{HL}=3} < \text{MSE}_{\text{HL}=2}$).

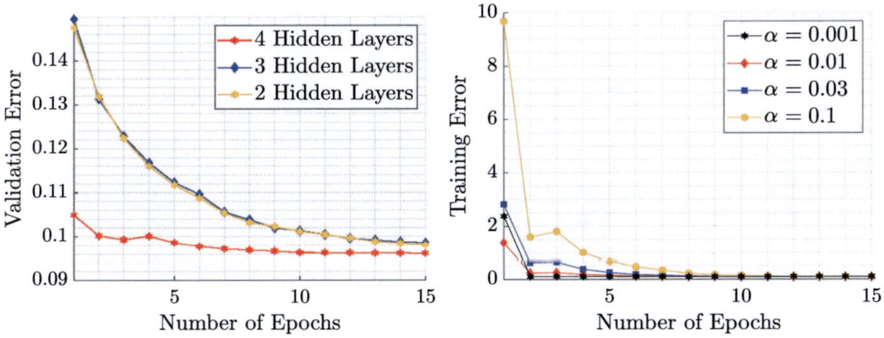

(a) (b)

Fig. 6.4 MSE loss vs. the number of epochs for different HLs: (**a**) Validation error. (**b**) Training error

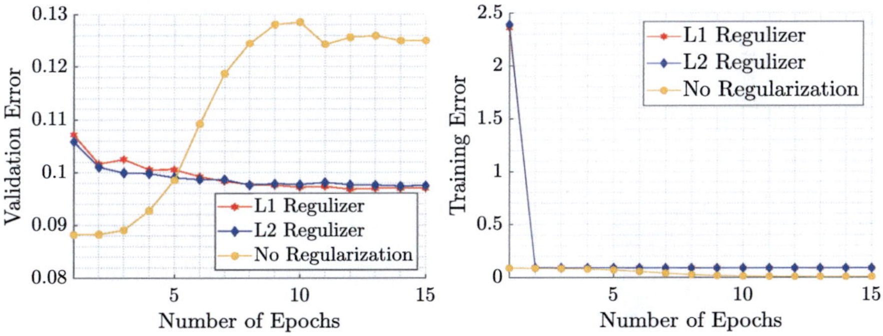

(a) (b)

Fig. 6.5 MSE loss vs. number of epochs for regularization: (**a**) Validation error. (**b**) Training error

To prevent overfitting of the training data, our proposed DNN architecture uses regularization techniques. Figure 6.5 compares the MSE performance for training and validation datasets in three scenarios: (1) no regularization; (2) L1 regularization; and (3) L2 regularization. Both L1 and L2 regularizations help avoid overfitting and enable the model to predict power and location values closer to an optimal solution on test data.[1] Another important factor in optimizing the DNN architecture is selecting an appropriate batch size (B_S), which can have a significant impact on memory usage, convergence speed, training stability, and model generalization. In Fig. 6.6a, we compare the MSE performance of the validation dataset for $B_S = \{2, 4, 16, 32, 64\}$. Our results indicate that $B_S = 4$, 16, or 32 can provide lower

[1] When there is no regularization, then the DNN architecture gives very low MSE for training dataset, but its performance on validation dataset degrades after a few epochs.

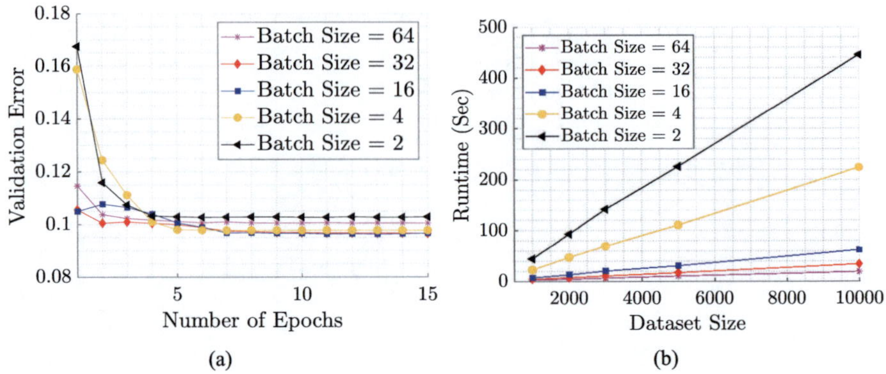

Fig. 6.6 MSE loss vs. number of epochs for different batch sizes: (**a**) Validation error. (**b**) Runtime

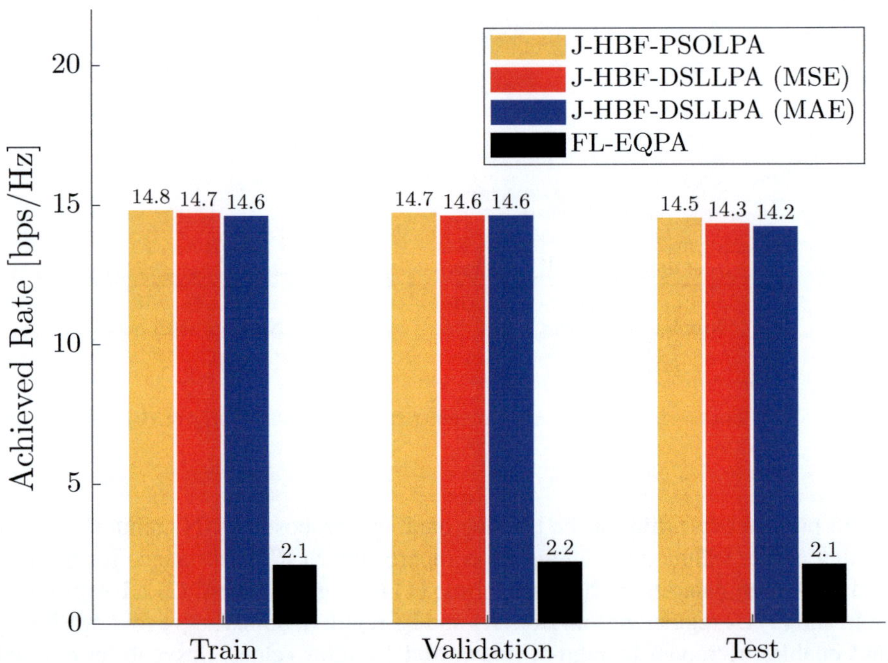

Fig. 6.7 Total AR evaluation on training, validation, and test datasets for J-HBF-PSOLPA and J-HBF-DSLLPA

MSE, but the runtime for $B_S = 4$ and 16 can be significantly longer than $B_s = 32$, especially for larger datasets with 100,000 realizations. Therefore, we use $B_S = 32$, which can provide a lower MSE with low computational complexity.

Figure 6.7 compares the total AR of J-HBF-PSOLPA with low-complexity DSL solutions, namely, J-HBF-DSLLPA with MSE, and J-HBF-DSLLPA with MAE, for $G = 1$, $K = 3$ at $P_T = 20$ dBm. We provide AR performance evaluations for training, validation, and test datasets. For benchmark comparison, we compare the proposed DSL-based solutions with J-HBF-PSOLPA and J-HBF-FL-EQPA (i.e., no optimization). The numerical results show that both DSL-based solutions can achieve AR close to the PSO-based solution for all datasets and outperform J-HBF-FL-EQPA. For instance, J-HBF-DSLLPA with MSE and MAE can provide the total AR of 14.3 and 14.2 bps/Hz, which is 98.6% and 97.4% of the total capacity achieved by J-HBF-PSOLPA, indicating a performance difference of only 1.3% and 2.06%, respectively. Furthermore, compared to J-HBF-FL-EQPA, both DSL-based solutions show a more than four-fold increase in AR.

6.4.2 *Complexity Analysis*

In this section, we compare the computational complexity of several proposed solutions for UAV-assisted MU-mMIMO communications systems. Specifically, we compare: (1) J-HBF-PSOPA-FL (discussed in Sect. 4.3.3.1); (2) J-HBF-PSOL-EQPA (presented in Sect. 4.3.3.2); (3) J-HBF-PSOLPA; and (4) J-HBF-DSLLPA. In Fig. 6.8, we present the runtime results for different numbers of users for $G = 2$,

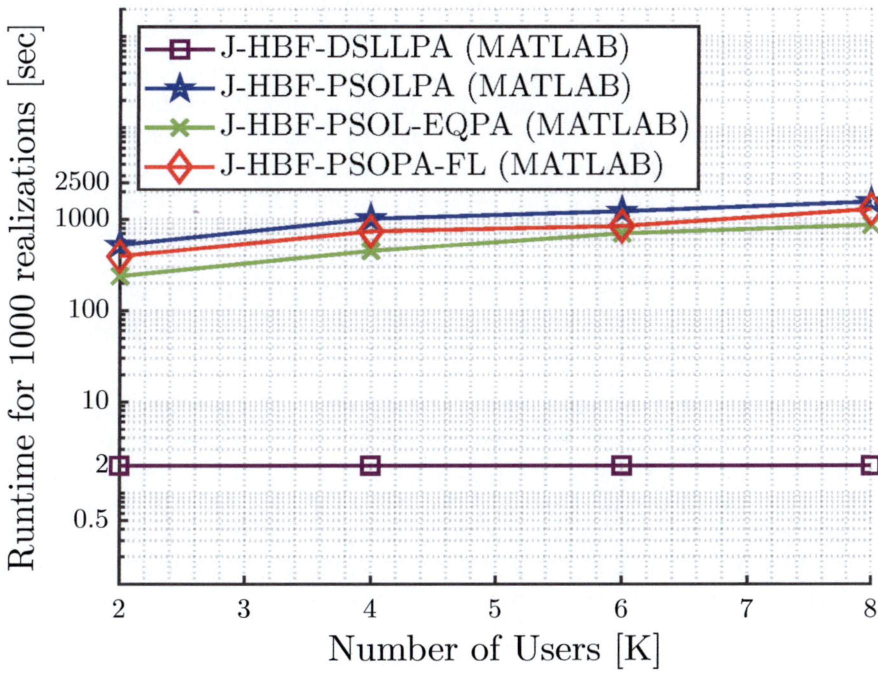

Fig. 6.8 Runtime comparison of proposed HBF solutions for different numbers of users

where $K = \{2, 4, 6, 8\}$. We provide the runtime for 1000 network realizations for test data. It is noteworthy to mention that the offline trained DNN architecture forJ-HBF-DSLLPA algorithm runs on MATLAB[2]. The proposed J-HBF-DSLLPA outperforms the computational complex J-HBF-PSOLPA, J-HBF-PSOL-EQPA, and J-HBF-PSOPA-FL algorithm by significantly reducing the runtime. For example, for a small number of users ($K = 2$), J-HBF-PSOLPA, J-HBF-PSOL-EQPA, and J-HBF-PSOPA-FL take approximately 489, 234, and 376 sec to predict $K+2$ values, whereas with J-HBF-DSLLPA, only about 2 sec are enough to predict the optimal UAV location and PA values. This means that the proposed J-HBF-DSLLPA with MSE can take only 0.4%–0.8% of the runtime of different PSO-based algorithmic solutions without impacting the total AR performance. When the number of users is increased to $K = 8$, then J-HBF-PSOLPA, J-HBF-PSOL-EQPA, and J-HBF-PSOPA-FL can take around 1725, 976, and 1453 sec, representing an increase of approximately 252%, 317%, and 287%, respectively, whereas with the proposed J-HBF-DSLLPA, the predictions take only about 2.2 sec, which is $\approx$0.127% of runtime of J-HBF-PSOLPA. Then, we consider the computational complexity of the proposed angular-based HBF design. For the RF, beamformers $\mathbf{F}_b$ and $\mathbf{F}_{u,r}$ require $O(N_T)$ and $O(N_r)$ operations, respectively. Meanwhile, for $\mathbf{F}_{u,t}$, the complexity is $O(GN_t)$ to serve G different user groups. For BB stages $\mathbf{B}_b$ and $\mathbf{B}_{u,r}$, we require $O(N^3_{RF_b})$ and $O(N^3_{RF_u})$ operations, respectively. For $\mathbf{B}_{u,t}$, the complexity is $O(N^3_{RF_u} + KN^2_{RF_u})$, where $O(KN^2_{RF_u})$ is for computation of $\mathcal{H}_2^H\mathcal{H}$ and $O(N^3_{RF_u})$ is for matrix inversion. Overall, the HBF design for BS and UAV requires $O(N_T + N_r + GN_t + N^3_{RF_b} + N^3_{RF_u} + KN^2_{RF_u})$ operations. Let us consider the J-HBF-PSOPA-FL algorithm, which requires $O(Z_p)$ to initialize particles and $O(K^2Z_p)$ operations to find K global solutions. Then, for a total of T PSO iterations, we need $O(TZ_p(N^3_{RF_u} + KN^2_{RF_u}))$. Ignoring the small computational terms (for instance, matrix multiplication, matrix inversion), the overall computational complexity of J-HBF-PSOPA-FL is $O(TZ_p(N^3_{RF_u} + KN^2_{RF_u}))$. Similarly, the complexities of J-HBF-PSOL-EQPA and J-HBF-PSOLPA are computed as $O(TZ_p(N_T+N_r+GN_t+N^3_{RF_b}+N^3_{RF_u}+KN^2_{RF_u})$ and $O(TZ_p(N_T+N_r+GN_t+N^3_{RF_b}+(N^3_{RF_u}+KN^2_{RF_u})^2)$, respectively [12]. Now, let us discuss the proposed DNN architecture with the hyper-parameters provided in Table 6.2. The complexity of the training phase can be calculated as follows:

- Forward Pass: (1) The *input layer* has $(2N_t+2N_{RFu}+2)K$ neurons, and computing the output of the input layer requires $O((2N_t + 2NRF_u + 2)K)$ operations; (2) there are four HL with 1024, 512, 256 and 128 neurons, respectively, and the total complexity of computing the output of each HL can be computed as follows: $O(1024)$ (layer 1), $O(512)$ (layer 2), $O(256)$ (layer 3), $O(128)$ (layer 4). Therefore, the total complexity of computing the output of all HL is $O(1024+512+256+128) = O(1920)$; (3) the *output layer* has $K+2$ neurons, so

[2] For the MATLAB runtime results, we implement both PSO-based J-HBF-PSOLPA and DSL-based J-HBF-DSLLPA via a PC with Intel Core(TM) i7-4770 CPU @ 3.4 GHz and 32 GB RAM.

the complexity of final layer is $\mathcal{O}(K+2)$. The overall complexity of the forward pass is $\mathcal{O}((2N_t+2N_{RFu}+2)K+3840+K+2)=\mathcal{O}((2N_t+2NRF_u+2)K)$.
- Backward Pass: The backward propagation involves computing gradients of the loss function with respect to the weights and biases of the DNN. The overall complexity of the backward pass is $\mathcal{O}(N_{\text{params}})$, where N_{params} is the total number of parameters in the DNN.[3]
- Assuming a batch size of 32 and 15 epochs, the total number of forward and backward passes would be $N_{\text{batch}} \cdot N_{\text{epochs}} = 480$. Therefore, the total computational complexity of the DNN during the training phase can be approximated as $\mathcal{O}(N_{\text{params}} N_{\text{batch}} N_{\text{epochs}}) \approx \mathcal{O}(3.42 \cdot 10^{11})$.

Compared to different proposed solutions using PSO, the DNN architecture has a higher computational complexity during the training phase; however, once fully trained, the proposed DNN architecture requires only $\mathcal{O}(1)$ operations for predicting $K+2$ output values for an input feature vector of size $(2N_t+2N_{RF_u}+2)K$. This is because the complexity is proportional to the number of layers and neurons in each layer, which is a constant value for a trained DNN. Thus, the proposed J-HBF-DSLLPA offers a low-complexity solution for real-time applications in UAV-assisted MU-mMIMO communications systems.

6.5 Concluding Remarks

In this chapter, a UAV-assisted MU-mMIMO communications system has been considered, where the UAVs operates as a DF relay between the BS and multiple users/devices. The problem of jointly designing the HBF stages for BS and UAV, together with UAV deployment and optimal PA to multiple users, is taken into consideration. For this challenging non-convex problem, we have proposed a novel DSL-based low-complexity J-HBF-DSLLPA to overcome the high computational complexity of the PSO-based solution discussed in Sect. 4.3.3.3. The results illustrate that the proposed J-HBF-DSLLPA can closely approach the capacity of the PSO-based solution, namely J-HBF-PSOLPA, and greatly reduce the runtime by 99%, which makes the implementation suitable for real-time online applications in UAV-assisted MU-mMIMO communication systems.

[3] The total number of parameters in the DNN can be calculated as follows: (i) input layer to HL 1 : $(2N_t+2N_{RF_u}+2)K \times 1024$, (ii) HL 1 to HL 2 : 1024×512, (iii) HL 2 to HL 3 : 512×256, (iv) HL 3 to HL 4 : 256×128, (v) HL 4 to output layer: 128.

Appendix: Regularization

In this chapter, we used L1 and L2 regularization techniques to prevent overfitting of the dataset. Overfitting is a condition in which the DL model has very good performance on training set (i.e., training error close to zero), while having poor performance on validation set (i.e., high validation error). In this appendix, we formulate the MSE loss function, while considering L1 and L2 regularizations. Considering L1 regularization, the MSE loss function in (6.12) will be modified as

$$\mathcal{L}_{\mathrm{MSE_{L1}}} = \underbrace{\frac{1}{B_S K}\sum_{i=1}^{B_S}\sum_{k=1}^{K}(\bar{p}_{k,i} - \hat{p}_{k,i})^2 + \frac{1}{B_S}\sum_{i=1}^{B_S}(\bar{x}_u - \hat{x}_u)^2 + \frac{1}{B_S}\sum_{i=1}^{B_S}(\bar{y}_u - \hat{y}_u)^2}_{\mathcal{L}_{\mathrm{MSE}} \text{ in (6.12)}}$$
$$+ \underbrace{\lambda_{\mathrm{L1}}\sum_{l=1}^{4}||\mathbf{W}_l||_1}_{\text{L1 term}}, \tag{6.14}$$

where λ_{L1} denotes the L1 regularization parameter and $||.||_1$ represents the element-wise 1-norm. Similarly, the MSE loss function with L2 regularization can be written as

$$\mathcal{L}_{\mathrm{MSE_{L2}}} = \underbrace{\frac{1}{B_S K}\sum_{i=1}^{B_S}\sum_{k=1}^{K}(\bar{p}_{k,i} - \hat{p}_{k,i})^2 + \frac{1}{B_S}\sum_{i=1}^{B_S}(\bar{x}_u - \hat{x}_u)^2 + \frac{1}{B_S}\sum_{i=1}^{B_S}(\bar{y}_u - \hat{y}_u)^2}_{\mathcal{L}_{\mathrm{MSE}} \text{ in (6.12)}}$$
$$+ \underbrace{\lambda_{\mathrm{L2}}\sum_{l=1}^{4}||\mathbf{W}_l||_F^2}_{\text{L1 term}}. \tag{6.15}$$

Here, λ_{L2} denotes the L2 regularization parameter. The key idea in both regularization methods is to penalize large weights by adding a term to the loss function that grows with each $\mathbf{W}_l$. In other words, increasing any weight $\mathbf{W}_l$ increases the overall loss; therefore, when we minimize the modified objectives in (6.14) and (6.15), the optimizer is driven to choose smaller $\mathbf{W}_l$ values. As a result, the learned weights tend to be closer to zero, reducing model complexity and making the solution less prone to overfitting.

While we used L1 and L2 regularization techniques in this book, other regularization methods such as Elastic Net [13], Dropout [14], and Early Stopping [15] can be considered for overfitting problems.

References

1. J. Chen, U. Yatnalli, D. Gesbert, Learning radio maps for UAV-aided wireless networks: a segmented regression approach, in *Proceedings of 2017 IEEE International Conference on Communication (ICC)* (2017), pp. 1–6
2. Q. Zhang, W. Saad, M. Bennis, X. Lu, M. Debbah, W. Zuo, Predictive deployment of UAV base stations in wireless networks: Machine learning meets contract theory. IEEE Trans. Wirel. Commun. **20**(1), 637–652 (2021)
3. K. Xiao, J. Zhao, Y. He, S. Yu, Trajectory prediction of UAV in Smart City using recurrent neural networks, in *Proceedings of 2019 International Conference on Communication (ICC)* (2019), pp. 1–6
4. J. Kwak, Y. Sung, Autoencoder-based candidate waypoint generation method for autonomous flight of multi-unmanned aerial vehicles. Adv. Mech. Eng. **11**(6), 1687814019856772 (2019)
5. X. Zhu, F. Qi, Y. Feng, Deep-learning-based multiple beamforming for 5G UAV IoT networks. IEEE Netw. **34**(5), 32–38 (2020)
6. F. Jiang, K. Wang, L. Dong, C. Pan, W. Xu, K. Yang, Deep-learning-based joint resource scheduling algorithms for hybrid MEC networks. IEEE Internet Things J. **7**(7), 6252–6265 (2020)
7. Y. Pan, Y. Yang, W. Li, A deep learning trained by genetic algorithm to improve the efficiency of path planning for data collection with multi-UAV. IEEE Access **9**, 7994–8005 (2021)
8. A. Koc, M. Wang, T. Le-Ngoc, Deep learning based multi-user power allocation and hybrid precoding in massive MIMO systems, in *Proceedings of the IEEE International Conference on Communication (ICC)* (2022), pp. 5487–5492
9. M.A. Nielsen, *Neural Networks and Deep Learning*, vol. 25 (Determination Press San Francisco, 2015)
10. T. Developers, *Tensorflow* (Zenodo, 2021)
11. A. Koc, A. Masmoudi, T. Le-Ngoc, 3D angular-based hybrid precoding and user grouping for uniform rectangular arrays in massive MU-MIMO systems. IEEE Access **8**, 84689–84712 (2020)
12. M. Mahmood, M. Ghadaksaz, A. Koc, T. Le-Ngoc, Deep learning meets swarm intelligence for UAV-assisted IoT coverage in massive MIMO. IEEE Internet Things J. **11**(5), 7679–7696 (2024)
13. H. Zou, T. Hastie, Regularization and variable selection via the elastic net. J. Roy. Stat. Soc. Ser. B (Statistical Methodology) **67**(2), 301–320 (2005)
14. P. Baldi, P. J. Sadowski, Understanding dropout, in *Advances in Neural Information Processing Systems*, vol. 26 (2013)
15. L. Prechelt, Early stopping-but when? in *Neural Networks: Tricks of the trade* (Springer, Berlin, 2002), pp. 55–69

Chapter 7
UAV Deployment Based on DRL in Dynamic Communications Networks

7.1 Introduction

The proposed DSL method shows promising performance in achieving a satisfactory AR and reducing runtime. However, it struggles in more complex scenarios where users have diverse distributions or in dynamic environments where users may change locations frequently. A potential solution is to train the DSL model on larger, more generalized datasets. However, this approach is impractical due to the immense computational resources required to generate such datasets, and even then, it may fail to encompass all possible scenarios. These limitations of DSL motivate us to explore RL as a more effective optimization strategy within ML.

As a cornerstone of AI, RL has been extensively researched in wireless communications and UAV applications [1–3]. RL is a decision-making approach that emphasizes learning through interaction with an environment. Within the spectrum of RL methods, the DDPG has emerged as a notable technique [4]. DDPG is particularly adept at handling continuous action spaces, which are common in real-world scenarios. This capability makes DDPG highly relevant for complex, dynamic environments where actions need to be precise and varied, as is often the case with UAV operations.

The deployment of the UAV plays an important role in enhancing the performance of UAV-assisted wireless communications systems, where different RL-based and DRL-based solutions were reviewed in Sect. 2.5.3 for UAV positioning [5–15]. In this chapter, we propose a joint HBF and UAV deployment framework using the DDPG-based algorithmic solution to maximize AR in dynamic MU-mMIMO communications systems. In particular, the RF beamforming stages for BS and UAV are designed based on the slow time-varying angular information, and BB stages are formulated using the reduced-dimensional effective channel matrices. Then, a novel DDPG-based algorithmic solution is proposed for UAV

T. Le-Ngoc et al., *AI-Enabled UAV-Assisted Massive MIMO*, Wireless Networks,
https://doi.org/10.1007/978-3-032-07872-8_7

deployment with a primary objective to not only maximize the overall AR in MU-mMIMO communications systems but also to significantly reduce computational complexity, particularly the runtime, compared to PSO as an NI optimization method. It is worthwhile to mention that the proposed DDPG algorithm exhibits a notable advantage in dynamic scenarios. The knowledge gained from training on the initial user's position is effectively transferred to subsequent positions, resulting in reduced learning time—akin to transfer learning. The illustrative results demonstrate the effectiveness of the proposed DDPG-based deployment scheme, achieving a runtime reduction of up to 31.5% compared to NI-based solutions. Additionally, the findings highlight the limitations of the DSL-based approach, particularly in more complex and dynamic environments [16].

The rest of this chapter is organized in the following manner. Section 7.2 defines the system and channel model for UAV relay MU-mMIMO systems. In Sect. 7.3, we introduce the joint HBF and DDPG-based UAV deployment framework. Section 7.4 presents the illustrative results. Finally, Sect. 7.5 concludes the chapter.

7.2 System and Channel Model

In this chapter, we extend the DF UAV relay system model presented in Sect. 4.2.1 by introducing mobility for the users/devices, allowing them to change their locations dynamically as depicted in Fig. 7.1. Similarly, we consider a BS equipped with N_T antennas, a UAV relay with N_r antennas for receiving and N_t antennas for sending information to K single-antenna users. Considering HBF for both BS and UAV, $\mathbf{B}_b \in \mathbb{C}^{N_{RF_b} \times K}$ and $\mathbf{F}_b \in \mathbb{C}^{N_T \times N_{RF_b}}$ denote the BB stage and RF stage at the BS, respectively, where N_{RF_b} is the number of RF chains. Similarly,

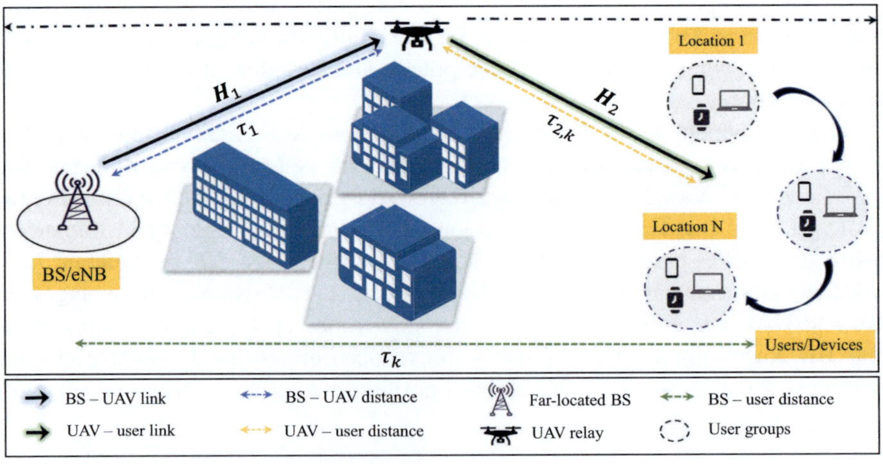

Fig. 7.1 UAV-assisted mmWave MU-mMIMO dynamic environment

$\mathbf{B}_{u,r} \in \mathbb{C}^{K \times N_{RFu}}$ and $\mathbf{F}_{u,r} \in \mathbb{C}^{N_{RFu} \times N_r}$ represent the UAV's receive BB and RF stage, respectively, while $\mathbf{B}_{u,t} = [\mathbf{b}_{u,t,1}, \cdots, \mathbf{b}_{u,t,K}] \in \mathbb{C}^{K \times N_t}$ and $\mathbf{F}_{u,t} = [\mathbf{f}_{u,t,1} \cdots, \mathbf{f}_{u,t,N_{RF_u}}] \in \mathbb{C}^{N_t \times N_{RFu}}$ are the UAV's transmit BB and RF stage, respectively. Additionally, $\mathbf{H}_1 \in \mathbb{C}^{N_r \times N_T}$ represents the channel matrix between the BS and UAV, while $\mathbf{H}_2 \in \mathbb{C}^{K \times N_t}$ denotes the channel between the UAV and the users. In this case, considering the environment noise follows the distribution $\mathcal{CN}(\mathbf{0}, \sigma_n^2)$, the total AR between the BS and UAV can be written as

$$\mathrm{R}_1(\mathbf{F}_b, \mathbf{B}_b, \mathbf{F}_{u,r}, \mathbf{B}_{u,r}) = \log_2 |\mathbf{I}_K + \mathbf{Q}_1^{-1} \mathbf{B}_{u,r} \mathcal{H}_1 \mathbf{B}_b \mathbf{B}_b^H \mathcal{H}_1^H \mathbf{B}_{u,r}^H|, \tag{7.1}$$

where $\mathbf{Q}_1^{-1} = (\sigma_n^2 \mathbf{B}_{u,r} \mathbf{F}_{u,r})^{-1} \mathbf{F}_{u,r}^H \mathbf{B}_{u,r}^H$ and $\mathcal{H}_1 = \mathbf{F}_{u,r} \mathbf{H}_1 \mathbf{F}_b$. Then, considering a single-user group (i.e., $G = 1$) and equal PA for the users, the SINR equation (4.8) in Sect. 4.2.1 can be simplified as

$$\mathrm{SINR}_k = \frac{|\mathbf{h}_{2,k}^H \mathbf{F}_{u,t} \mathbf{b}_{u,t,k}|^2}{\sum\limits_{\hat{k} \neq k}^{K} |\mathbf{h}_{2,\hat{k}}^H \mathbf{F}_{u,t} \mathbf{b}_{u,t,\hat{k}}|^2 + \sigma_n^2}. \tag{7.2}$$

Utilizing the instantaneous SINR, the ergodic sum rate capacity of the UAV-users link is described as

$$\mathrm{R}_2(\mathbf{F}_{u,t}, \mathbf{B}_{u,t}, x_u, y_u) = \mathbb{E}\left\{ \sum_{k=1}^{K} \mathbb{E}\left[\log_2(1 + \mathrm{SINR}_k)\right] \right\}, \tag{7.3}$$

where x_u (y_u) denotes the UAV location in the x-axis (y-axis).

To account for the effects of multipath fading and shadowing, we adopt the mmWave channel model for both the BS-UAV and UAV-user links, utilizing the Saleh-Valenzuela channel model [17]. The derivation of the channel matrices $\mathbf{H}_1$ and $\mathbf{H}_2$ is provided in Sect. 4.2.2, and the same approach is applied in this chapter.

7.2.1 Problem Formulation

The total AR can be maximized by the joint optimization of the beamforming stages $\mathbf{F}_b$, $\mathbf{B}_b$, $\mathbf{F}_{u,t}$, $\mathbf{F}_{u,r}$, $\mathbf{B}_{u,t}$, and $\mathbf{B}_{u,r}$ with optimal UAV positioning $\mathbf{x}_u = [x_u, y_u]^T \in \mathbb{R}^2$. Here, $\mathbf{x}_u$ represents the 2-D UAV deployment in a given flying span z_u.

Subsequently, the problem can be formulated as follows:

$$\begin{aligned}
&\max_{\{\mathbf{F}_b,\mathbf{B}_b,\mathbf{F}_{u,r},\mathbf{B}_{u,r},\mathbf{F}_{u,t},\mathbf{B}_{u,t},\mathbf{x}_u\}} \quad \mathrm{R_T} \\
&\text{s.t.} \quad C_1 : |\mathbf{F}_b(i,j)| = \frac{1}{\sqrt{N_T}} \quad \forall i,j, \\
&\qquad C_2 : |\mathbf{F}_{u,r}(i,j)| = \frac{1}{\sqrt{N_r}}, \quad |\mathbf{F}_{u,t}(i,j)| = \frac{1}{\sqrt{N_t}} \quad \forall i,j, \\
&\qquad C_3 : \mathbb{E}\{\|\mathbf{s}_1\|_2^2\} \le P_T, \\
&\qquad C_4 : \mathbb{E}\{\|\mathbf{s}_2\|_2^2\} \le \sum_{k=1}^{K} \mathbf{b}_{u,t,k}^H \mathbf{F}_{u,t}^H \mathbf{F}_{u,t} \mathbf{F}_{u,t} \mathbf{b}_{u,t,k} \le P_T, \\
&\qquad C_5 : x_{\min} \le x_u \le x_{\max}, \quad y_{\min} \le y_u \le y_{\max},
\end{aligned} \tag{7.4}$$

where $\mathrm{R_T} = \frac{1}{2}\min(\mathrm{R}_1, \mathrm{R}_2)$ is the total transmission rate from BS to multiple devices under DF protocol, C_1 and C_2 refer to the CM constraint for UAV and BS, respectively, C_3 and C_4 represent the transmit power constraint for BS and UAV, respectively, and C_5 implies UAV deployment within the given flying span. Here, $[x_{\min}, x_{\max}]$ ($[y_{\min}, y_{\max}]$) represents the UAV deployment range in the x-axis (y-axis).

7.3 Joint HBF and UAV Deployment Using DDPG

For given RF stages $\mathbf{F}_b$, $\mathbf{F}_{u,r}$, and $\mathbf{F}_{u,t}$ (designed in Sect. 4.3.1) as well as the BB stages $\mathbf{B}_b$, $\mathbf{B}_{u,r}$, and $\mathbf{B}_{u,t}$ (developed in Sect. 4.3.2), the optimization problem in (7.4) can be simplified as

$$\begin{aligned}
&\max_{\{\mathbf{x}_u\}} \quad \mathrm{R_T}\left(\mathbf{F}_b, \mathbf{B}_b, \mathbf{F}_{u,r}, \mathbf{B}_{u,r}, \mathbf{F}_{u,t}, \mathbf{B}_{u,t}, \mathbf{x}_u\right) \\
&\text{s.t.} \quad C_5 : x_{\min} \le x_u \le x_{\max}, \quad y_{\min} \le y_u \le y_{\max},
\end{aligned} \tag{7.5}$$

where at each deployment point the BB stages are first updated according to the new CSI, while RF stages remain constant as they have been developed based on the slow time-varying angular information. To tackle the problem stated in (7.5), we propose a DDPG-based solution, utilizing a DRL approach with a continuous action space to solve the UAV deployment problem. We begin by outlining the key principles of the DDPG algorithm, followed by the development of a DDPG-based framework for the UAV deployment problem.

7.3.1 DDPG: Preliminaries

DDPG is an advanced DRL algorithm that merges features of deep Q-network (DQN) with policy gradient techniques. While DQN is optimized for discrete action spaces, DDPG is specifically designed for continuous action spaces, which are prevalent in practical applications. This makes DDPG highly effective for handling complex tasks that necessitate a variety of continuous action values, thereby improving its performance in varied and dynamic environments.

DDPG algorithm employs an actor-critic approach [4], as depicted in Fig. 7.2. Specifically, at each time step t, the actor network generates a deterministic action $\mathbf{a}_t$ based on the input state $\mathbf{s}_t$, while the critic network estimates the expected return (value) of executing the action $\mathbf{a}_t$ in state $\mathbf{s}_t$, denoted as $\mu(\mathbf{s}_t|\mathbf{W}^{\mu})$ and $Q(\mathbf{s}_t, \mathbf{a}_t|\mathbf{W}^{Q})$, respectively. Here, $\mathbf{W}^{\mu}$ represents the weights of the actor network, while $\mathbf{W}^{Q}$ indicates the weights of the critic network. DDPG also uses target actor and target critic networks, denoted as $\mu'(\mathbf{s}_t|\mathbf{W}^{\mu'})$ and $Q'(\mathbf{s}_t, \mathbf{a}_t|\mathbf{W}^{Q'})$, respectively, that are the slow-paced versions of main actor and critic networks to ensure a more stable training process. In this context, $\mathbf{W}^{\mu'}$ is the weight of the target actor, and $\mathbf{W}^{Q'}$ refers to the weight of the target critic network. Additionally, to disrupt the

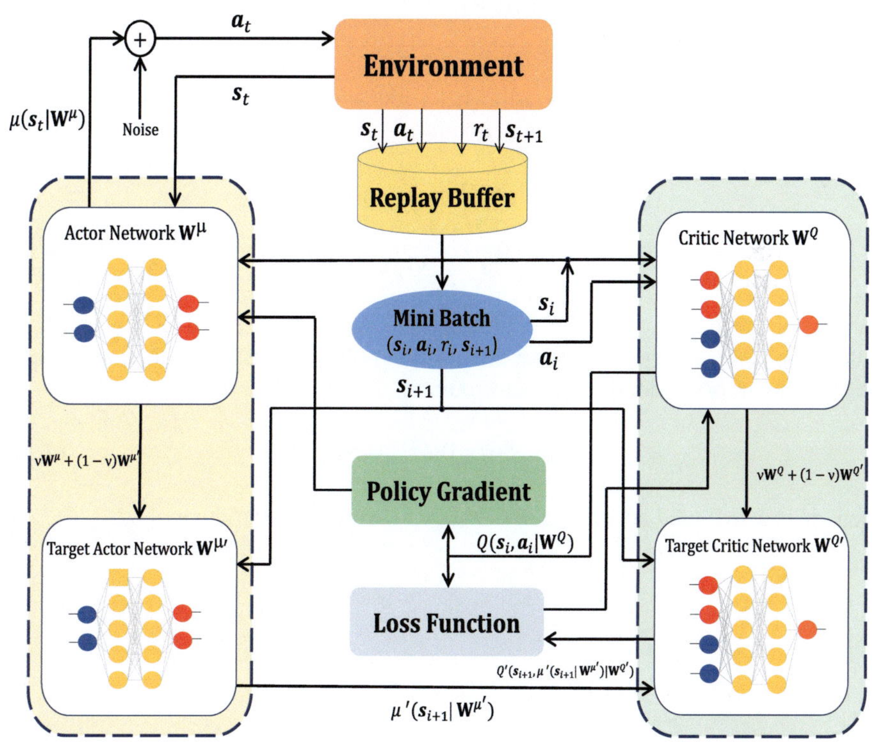

Fig. 7.2 Block diagram of DDPG algorithm

correlation between consecutive experiences ($\mathbf{s}_t$, $\mathbf{a}_t$, r_t, $\mathbf{s}_{t+1}$), we employ a replay buffer D. Here, r_t is the reward received at time step t, and $\mathbf{s}_{t+1}$ denotes the state at the subsequent time step. These transitions are stored in the replay buffer. At each time step, the actor and critic networks are updated by sampling a minibatch of B_S transitions ($\mathbf{s}_i$, $\mathbf{a}_i$, r_i, $\mathbf{s}_{i+1}$) uniformly from this buffer, where the subscript i represents the ith sample out of B_S samples. In this case, the weights of actor and critic networks can be updated by minimizing the MSE loss function as[1]

$$\mathcal{L}_{\text{MSE}} = \frac{1}{B_S}\sum_i (y_i - Q(\mathbf{s}_i, \mathbf{a}_i|\mathbf{W}^Q))^2, \tag{7.6}$$

where y_i can be expressed as

$$y_i = r_i + \gamma Q'(\mathbf{s}_{i+1}, \mu'(\mathbf{s}_{i+1}|\mathbf{W}^{\mu'})|\mathbf{W}^{Q'}). \tag{7.7}$$

Here, the discount factor $\gamma \in [0, 1]$ is introduced to prevent the accumulation of infinite rewards. Subsequently, the actor network is updated utilizing the policy gradient method as

$$\nabla_{\mathbf{W}^\mu} J \approx \frac{1}{B_S}\sum_i \nabla_a Q(\mathbf{s}, \mathbf{a}|\mathbf{W}^Q)|_{\mathbf{s}=\mathbf{s}_i, a=\mu_{\mathbf{s}_i}} \nabla_{\mathbf{W}^\mu}\mu(\mathbf{s}|\mathbf{W}^\mu)|_{\mathbf{s}_i}. \tag{7.8}$$

Afterward, the target actor and critic networks are updated using a soft update approach, as shown in the following equations:

$$\mathbf{W}^{Q'} \leftarrow \nu\mathbf{W}^Q + (1-\nu)\mathbf{W}^{Q'}, \tag{7.9}$$

$$\mathbf{W}^{\mu'} \leftarrow \nu\mathbf{W}^\mu + (1-\nu)\mathbf{W}^{\mu'}, \tag{7.10}$$

where ν is the target networks' learning rate. As is common in RL algorithms, exploration is necessary to discover the optimal policy. To facilitate this, zero-mean Gaussian noise $\mathcal{N}$ is added to the actor policy:

$$\mathbf{a}_t = \mu(\mathbf{s}_t|\mathbf{W}^\mu) + \mathcal{N}, \tag{7.11}$$

where $\mathcal{N}$ is distributed according to $\mathcal{CN}(0, \sigma_e^2)$ with noise variance of σ_e.

[1] More details can be found in appendix of this chapter.

7.3.2 *DDPG-Based UAV Deployment*

To apply the DDPG algorithm to our problem, we need to define appropriate states, actions, and rewards for the agent. Moreover, the design of the actor and critic network can significantly influence the performance of the algorithm. Thus, first, we introduce the suitable state, action, and reward for UAV deployment, and then, we discuss the configuration of the actor and critic network.

7.3.2.1 State

At each time step t, the UAV agent observes the state $\mathbf{s}_t$ as follows:

$$\mathbf{s}_t = [\hat{x}_{u,t}, \hat{y}_{u,t}]^T \in \mathbb{R}^2, \tag{7.12}$$

where $\hat{x}_{u,t} = \frac{x_{u,t}}{x_{\max}}$ and $\hat{y}_{u,t} = \frac{y_{u,t}}{y_{\max}}$ denote the normalized 2-D location of the UAV at time step t. Here, the UAV is deployed at a fixed height $z_{u,t}$.[2]

7.3.2.2 Action

At each time step t, the UAV agent's action $\mathbf{a}_t$ is defined as

$$\mathbf{a}_t = [a_{t,x}, a_{t,y}]^T \in \mathbb{R}^2, \quad \begin{cases} a_{t,x} \in [-a_{x,\max}, a_{x,\max}] \\ a_{t,y} \in [-a_{y,\max}, a_{y,\max}]. \end{cases} \tag{7.13}$$

Here, $a_{x,\max}$ ($a_{y,\max}$) represents the maximum movement step for the UAV on the x-axis (y-axis). A positive $a_{t,x}$ ($a_{t,y}$) indicates movement to the East (North), while a negative $a_{t,x}$ ($a_{t,y}$) indicates movement to the West (South).

7.3.2.3 Reward

The reward function conveys the task objectives to the agent. Properly designing the reward function is crucial as it shapes the learning process, influences convergence speed, and affects the overall effectiveness of the learned policy. We define the reward function r_t at time step t as

$$r_t = \begin{cases} \mathrm{R_T} \text{ for } \mathrm{R_T} \geq \Omega_0 \\ -1 \text{ for } \mathrm{R_T} < \Omega_0 \\ -5 \text{ for } x_{u,t} > x_{\max} \text{ or } y_{u,t} > y_{\max} \\ -5 \text{ for } x_{u,t} < x_{\min} \text{ or } y_{u,t} < y_{\min}, \end{cases} \tag{7.14}$$

[2] For simplicity, we consider a scenario of fixed UAV height. However, the proposed DDPG-based solution can be applied for 3-D UAV deployment, which is left as our future work.

where R_T denotes the AR as given in (7.5), Ω_0 is an adjustable threshold, $x_{\max}$ ($x_{\min}$) is the maximum (minimum) allowable position on the x-axis, and $y_{\max}$ ($y_{\min}$) is the maximum (minimum) permitted position on the y-axis. Negative rewards help the agent find the best deployment by penalizing the UAV for making poor actions. For instance, a penalty of -1 is given when the UAV explores an area with a low AR, which discourages the agent from focusing on less useful areas. Similarly, the UAV must stay within the deployment span. If it goes outside this area, a larger penalty of -5 is given to further discourage it from exploring beyond the permitted deployment span.[3]

7.3.2.4 Actor and Critic Networks

We use a fully connected DNN architecture with two HLs for both the actor and critic networks, as shown in Fig. 7.3. The actor network predicts suitable actions based on the input states, while the critic network determines the Q-value of the input action-state pair. Each HL of the actor network contains ζ_i^a neurons, with $i = \{1, 2\}$, and each HL of the critic network contains ζ_j^c neurons, with $j = \{1, 2\}$. Also, to perform nonlinear operations, we use the ReLU activation function (i.e., $f_r(z) = \max(0, z)$) in the HLs for both actor and critic networks to accelerate the learning process. For the actor network's output layer, we use the Tanh activation function (i.e., $f_t(z) = \frac{e^z - e^{-z}}{e^z + e^{-z}}$) to ensure the predicted actions are within $[-a_{x,\max}, a_{x,\max}]$ ($[-a_{y,\max}, a_{y,\max}]$) for the x-axis (y-axis). The critic network's output layer uses a

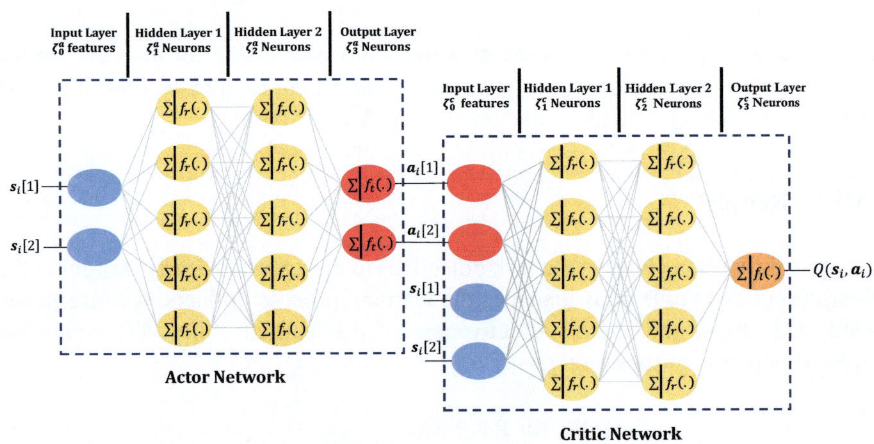

Fig. 7.3 (Target) actor and (target) critic DNN architecture

[3] Here, our objective is to find the optimal UAV location within a given deployment span, and thus we select arbitrary values. The selection and optimization of negative rewards is beyond the scope of this thesis.

Algorithm 7.1: DDPG-based UAV deployment

Randomly initialize actor $\mu(\mathbf{s}|\mathbf{W}^{\mu})$ and critic $Q(\mathbf{s}, \mathbf{a}|\mathbf{W}^{Q})$ networks with weights $\mathbf{W}^{\mu}$ and $\mathbf{W}^{Q}$.
Initialize target network Q' and μ' with weights $\mathbf{W}^{Q'} \leftarrow \mathbf{W}^{Q}, \mathbf{W}^{\mu'} \leftarrow \mathbf{W}^{\mu}$.
Initialize replay buffer D.
for *episode=1:M* **do**
    Receive initial observation state $\mathbf{s}_1$.
    for *t=1:T* **do**
        Select action $\mathbf{a}_t$ using (7.11).
        Execute action $\mathbf{a}_t$ and observe reward r_t and new state $\mathbf{s}_{t+1}$.
        Store transition $(\mathbf{s}_t, \mathbf{a}_t, r_t, \mathbf{s}_{t+1})$ in D.
        Sample a random minibatch of B_S transitions $(\mathbf{s}_i, \mathbf{a}_i, r_i, \mathbf{s}_{i+1})$ from D.
        Formulate y_i using (7.7).
        Update the critic by minimizing the loss function $\mathcal{L}_{\text{MSE}}$ in (7.6).
        Update the actor using the policy gradient in (7.8).
        Update the target networks via (7.9), (7.10).
    end
end

linear activation function (i.e., $f_l(z) = z$) due to the large or unbounded range of Q-values. The summary of the offline phase is provided in Algorithm 7.1.

7.4 Illustrative Results

In this section, we present illustrative results on the performance of the DDPG-UD algorithm for different scenarios. For benchmark comparison, we compare the proposed DDPG-UD with the following solutions: (1) PSO-UD (the details on the PSO-based UAV deployment can be found in Sect. 4.3.3.2); and (2) deep learning (DL)-based UAV deployment (DL-UD), (i.e., DSL approach). Table 7.1 outlines the simulation setup based on the 3-D microcell scenario [18] for the results discussed hereafter. First, we fine-tune the hyper-parameters for the DDPG-UD to achieve optimal results. We then evaluate the algorithm's performance in two distinct types of environments: (1) static (fixed user locations) environments and (2) dynamic (users changing locations) environments.

7.4.1 Hyper-parameter Tuning

To achieve optimal performance, we first fine-tuned the hyper-parameters for the DDPG-UD. In this case, we use $K = 4$ users whose locations are within $(x_k, y_k) \in [50, 100]$. Additionally, to ensure more reliable results, the learning

Table 7.1 Simulation parameters

Number of antennas		$(N_T, N_t, N_r) = 144$	
BS height	UAV height	10 m	20 m
UAV x-axis range	UAV y-axis range	$[x_{\min}, x_{\max}] =$ [0, 100] m	$[y_{\min}, y_{\max}] =$ [0, 100] m
UAV x-axis movement	UAV y-axis movement	$[a_{x,\min}, a_{x,\max}] =$ $[-1, 1]$ m	$[a_{y,\min}, a_{y,\max}] =$ $[-1, 1]$ m
# of paths	Path loss exponent	$L = 10$	3.6
Noise PSD	Reference path loss α	− 174 dBm/Hz	61.34 dB
Frequency	Channel bandwidth	28 GHz	100 MHz
Mean AAoD/AAoA (1st link)	Mean AAoD (2nd link)	120°	$\phi_g = 21° + 120°(g-1)$
Mean EAoD/EAoA (1st link)	Mean EAoD (2nd link)	60°	$\theta_g = 60°$
Azimuth/elevation angle spread	# of network realization	±10°	2000

curves are averaged over five independent runs, maintaining the same user locations for consistency.

Figure 7.4 illustrates the accumulated reward achieved by the agent over 500 episodes for various combinations of actor and critic learning rates $(\alpha, \beta) = \{(0.0001, 0.0002), (0.001, 0.002), (0.01, 0.02)\}$, paired with different target network learning rates $\nu = \{0.001, 0.01, 0.1\}$. The figure reveals that the best choice for the actor and critic learning rates is $(\alpha, \beta) = (0.001, 0.002)$ across all target network learning rates. Other combinations of α and β either converge to suboptimal solutions or fail to learn effectively. Additionally, it is evident that within the $(\alpha, \beta) = (0.001, 0.002)$ configuration, a target network learning rate of $\nu = 0.01$ yields the highest accumulated reward compared to $\nu = 0.1$ and $\nu = 0.001$. This underscores the critical importance of carefully selecting all learning rates—α, β, and ν—to achieve the best performance. Similarly, Fig. 7.5a compares the accumulated reward performance across different numbers of HLs in the actor-critic architecture. We evaluate the performance over 500 episodes for architectures with 1, 2, and 3 HLs. The results show that the 2 HLs architecture achieves the highest accumulated reward. While the 1 HL architecture performs worse than the 2-layer setup due to its limited capacity to capture the complexity of the problem, the 3 HLs architecture yields the lowest accumulated reward, indicating that its complexity is excessive for this particular problem, and it may require a large number of episodes to achieve better performance.

Another crucial factor in optimizing the DDPG algorithm is selecting the appropriate batch size, as it significantly impacts memory usage, convergence speed, training stability, and the model's generalization ability. In Fig. 7.5b, we compare the accumulated reward for batch size = {8, 16, 32, 64}. Our results show that batch sizes = 32 and batch size = 16 offer the best performance in terms of both stability and accumulated reward. However, batch size = 32 leads to faster

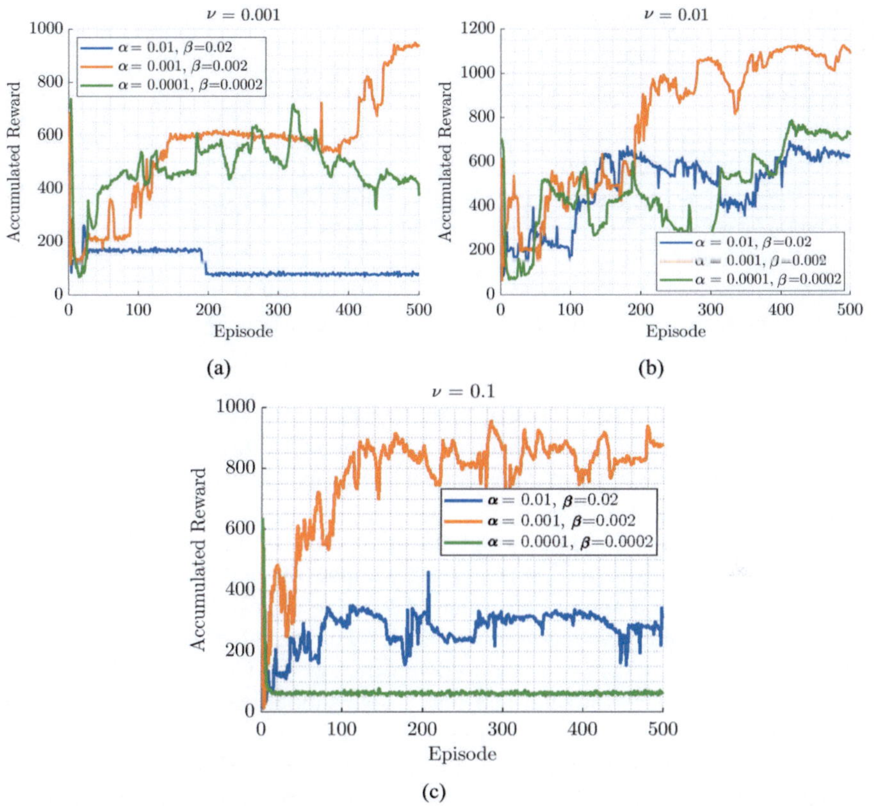

Fig. 7.4 Accumulated reward vs. number of episodes for: (**a**) $\nu = 0.001$, (**b**) $\nu = 0.01$, and (**c**) $\nu = 0.1$

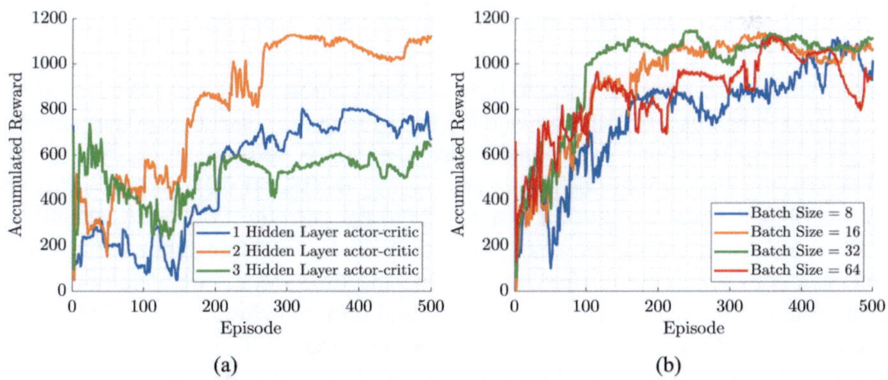

Fig. 7.5 Accumulated reward vs. number of episodes for: (**a**) different HLs and (**b**) different batch sizes

Table 7.2 Networks' parameters

(Target) actor network architecture			
Input shape	$\zeta_0^a = 2$	1st hidden layer	$\zeta_1^a = 20$
2nd hidden layer	$\zeta_2^a = 20$	Output layer	$\zeta_3^a = 2$
(Target) critic network architecture			
Input shape	$\zeta_0^c = 4$	1st hidden layer	$\zeta_1^c = 20$
2nd hidden layer	$\zeta_2^c = 20$	Output layer	$\zeta_3^c = 1$
Network parameters			
Reply buffer size	$D = 60000$	Critic learning rate	$\beta = 0.002$
Actor learning rate	$\alpha = 0.001$	Target networks learning rate	$\nu = 0.01$

convergence compared to batch size = 16. Consequently, we select a batch size = 32, as it delivers a strong accumulated reward, while ensuring quicker convergence. Table 7.2 outlines the hyper-parameters for the proposed actor-critic architecture, which is given in Fig. 7.3.

In addition to hyper-parameters and the architecture of the actor-critic networks, the reward function given in (7.14) can greatly impact the performance of the DDPG agent. Specifically, the appropriate value of variable Ω_0 can help the agent to find the optimal location in less number of episodes. Higher values of Ω_0 will make the reward more sparse, which can lead to divergence of the UAV, while lower values of Ω_0 will make the reward flat and may cause suboptimal convergence. Fig.7.6 shows the accumulated reward for three values of Ω_0. It can be seen that the agent achieves the highest return with $\Omega_0 = 5$ bps/Hz at $P_T = 30$ dBm. Here, the value of Ω_0 is dependent on the transmission power and needs to be tuned for each P_T.

7.4.2 *Static Environment (Fixed Users Locations)*

In this section, we consider users to have a fixed location and discuss two scenarios: (1) narrow-range user distribution and (2) wide-range user distribution. For narrow-range user distribution, we consider that the BS is located at $(x_b, y_b, z_b) = (0, 0, 10)$, the UAV initial position is $(x_u, y_u, z_u) = (50, 50, 20)$, and $K = 4$ users are distributed randomly at a far distance from the BS (i.e., $(x_k, y_k) \in [90, 100]$). UAV starts its initial location at $(50, 50, 20)$ at the beginning of each episode, and then it explores the environment during the time steps. For wide-range user distribution, we consider the same location and initial position for the UAV, while assuming that $K = 4$ users are randomly scattered with $(x_k, y_k) \in [50, 100]$. Figure 7.7 shows the achieved rates for DDPG-UD, PSO-UD, DL-UD, and fixed UAV deployment (FD) (i.e., no optimization). Numerical results reveal that the proposed DDPG-UD can achieve 98.63% of PSO-UD performance for narrow-range user distribution, while having a 16 times better performance than FD.

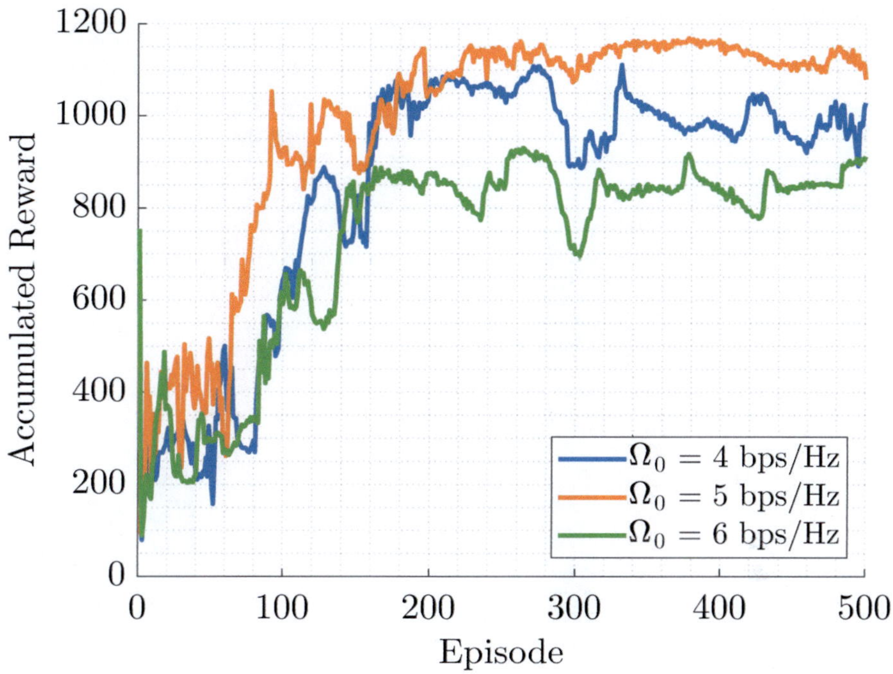

Fig. 7.6 Accumulated reward vs. number of episodes for different values of Ω_0

Furthermore, DDPG-UD can enhance the performance of PSO-UD by 2.23% for wide-range user distribution, while having 3.14 times better AR than the FD. Figure 7.7 also demonstrates that although DL-UD can have an acceptable performance when we have narrow-range user distribution, while achieving 94.09% AR of PSO-UD, it fails to find the optimal location for wide-range user distribution having only 36.94% AR of the DDPG-UD. This means that for more complex user distributions, DL-UD is not a promising solution. For the next step, we consider that users may change their location during the observation.

7.4.3 Dynamic Environment (Changing Users Locations)

In this section, we consider a more practical scenario where users can change their location during the training. This scenario represents dynamic, real-world environments. We assume that the BS is set at $(x_b, y_b, z_b) = [0, 0, 10]$. Then we

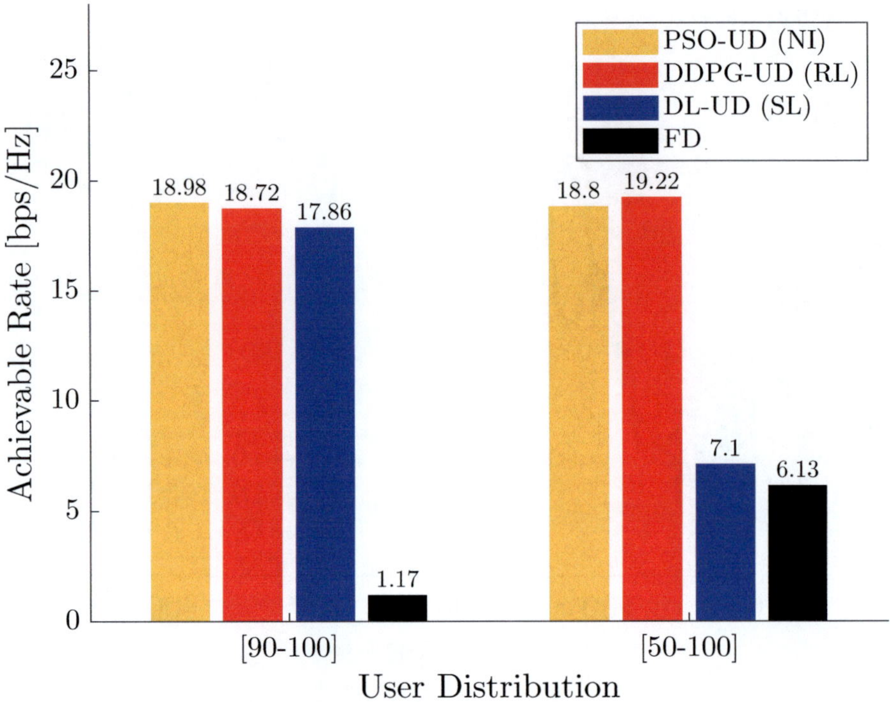

Fig. 7.7 Total AR for different user distributions

set the users' locations randomly at six different distributions $l \in \{l_1, l_2, \ldots, l_6\}$ as follows:

$$(x_{k,l}, y_{k,l}) = \begin{cases} x_{k,l} \in [60, 70], & y_{k,l} \in [60, 70] \text{ for } l_1 \\ x_{k,l} \in [60, 70], & y_{k,l} \in [70, 80] \text{ for } l_2 \\ x_{k,l} \in [70, 80], & y_{k,l} \in [80, 90] \text{ for } l_3 \\ x_{k,l} \in [80, 90], & y_{k,l} \in [80, 90] \text{ for } l_4 \\ x_{k,l} \in [80, 90], & y_{k,l} \in [70, 80] \text{ for } l_5 \\ x_{k,l} \in [80, 90], & y_{k,l} \in [60, 70] \text{ for } l_6 \end{cases}, \tag{7.15}$$

where $x_{k,l}$ and $y_{k,l}$ denote the kth user x-axis and y-axis ranges during lth distribution, respectively. Furthermore, UAV starts its initial location at $(x_u, y_u, z_u) = (50, 50, 20)$ and finds the optimal deployment $\mathbf{x}_u^{(1)} = \{x_u^{(1)}, y_u^{(1)}\}$ for l_1, which is now used as its initial location for the next user distribution (i.e., l_2). Figure 7.8 shows the accumulated reward and average accumulated reward for the DDPG agent during six different user locations. This figure shows that after the first user location, the DDPG agent maintains the reward and finds the optimal location in less number of episodes. The reason is that the pretrained networks from previous user locations contain information about the environment, thus helping the DDPG agent find the

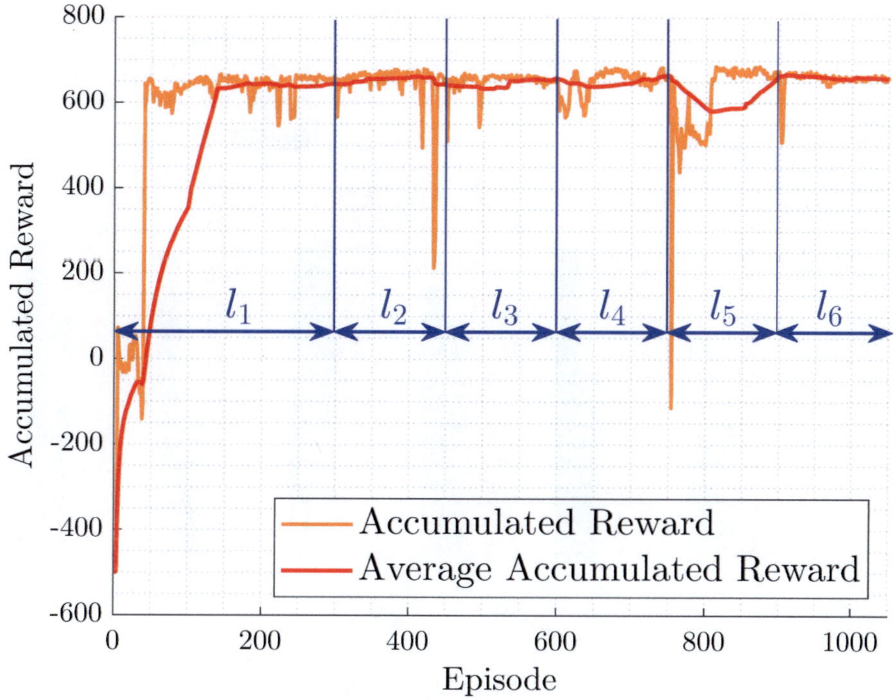

Fig. 7.8 Accumulated reward vs. episode for DDPG

optimal UAV location for changing user locations in a shorter time. Figure 7.9 shows the achieved rate for DDPG-UD, PSO-UD, DL-UD, and FD, which shows that DDPG-UD can achieve a performance close to PSO-UD. For instance, DDPG-UD accomplishes 99.62% of PSO-UD AR for l_3, and it provides 18.54 bps/Hz AR at l_6 and achieves 99.42% of PSO-UD performance. Additionally, the capacity is improved by 36.99% and 16.73% for FD for l_3 and l_6, respectively. Figure 7.9 also supports the previous deduction that DL-UD is not a promising solution for complex scenarios, while having only 63.84% and 66.65% of PSO-UD performance for l_3 and l_6, respectively.

Figure 7.10 displays the time comparison between DDPG-UD and PSO-UD. Here, we provide the runtime for a single-user location (i.e., $l \in l_1$), three different user locations ($l \in \{l_1, l_2, l_3\}$), and six distinct user locations ($l \in \{l_1, \cdots, l_6\}$). This figure shows that although for a single-user location, DDPG-UD takes 18.04% more than PSO-UD to find the optimal location, for multiple user locations, it takes 76.45% and 68.50% of PSO-UD runtime. This means that as the number of user locations increases, DDPG-UD will take less time to find the optimal location when compared to PSO-UD, which makes DDPG-UD an efficient solution for dynamic and fast-changing environments in MU-mMIMO communications systems.

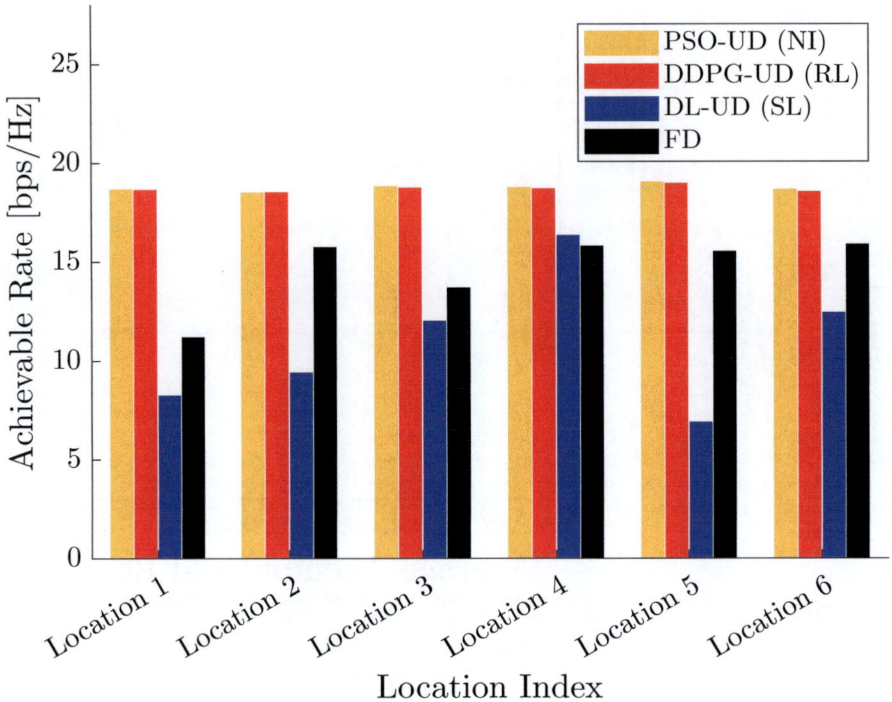

Fig. 7.9 Total AR in different locations

7.5 Concluding Remarks

In this chapter, a novel DDPG-UD and HBF technique has been proposed for AR maximization in dynamic MU-mMIMO communications systems. First, we introduce HBF for both the BS and UAV. Afterward, we apply the DDPG as a DRL approach to UAV deployment in dynamic environments. Illustrative results show that the proposed DDPG-UD closely approaches the optimal rate achieved by PSO-UD. On the other hand, DDPG-UD greatly reduces the runtime by 31.5%, which makes DDPG-UD a more appropriate solution for real-world dynamic applications in UAV-assisted mMIMO communications systems. Additionally, it is important to note that DL-UD, as an DSL method, struggles in more complex, dynamic scenarios. This limitation further highlights the effectiveness of the proposed DDPG-UD, which adapts well to dynamic settings.

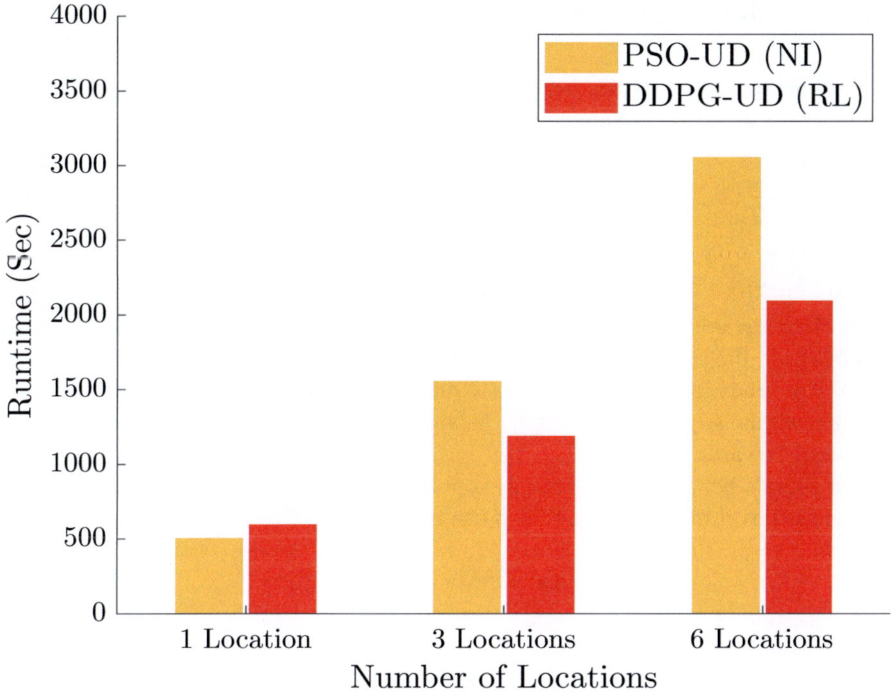

Fig. 7.10 Runtime for different numbers of locations

Appendix: Markov Decision Process

In Sect. 2.5.3, we introduce RL as a sequential decision-making process in which an agent interacts with an environment through various time steps. Specifically, we discussed that at each time step t, the agent receives a state $\mathbf{s}_t$, and accordingly, takes an action $\mathbf{a}_t$. Then, the agent receives a numerical reward r_{t+1}, a feedback from the environment, and the next state $\mathbf{s}_{t+1}$.

The above-mentioned agent-environment interaction is called Markov decision process. In other words, generally, all the previous states, not only state $\mathbf{s}_t$, will have impact on the next state $\mathbf{s}_{t+1}$. However, by considering the Markov property, we assume that only the previous state impacts the next state. This feature can be expressed mathematically as

$$\mathcal{P}\left(\mathbf{s}_{t+1}|\mathbf{s}_t\right) = \mathcal{P}\left(\mathbf{s}_{t+1}|\mathbf{s}_1, \mathbf{s}_2, \cdots, \mathbf{s}_t\right), \tag{7.16}$$

where $\mathcal{P}\left(.\right)$ denotes the probability function. Considering the Markov property, we can simplify the modeling of the RL problems.

Action-Value (Q) Function

The ultimate goal of RL is to maximize the cumulative reward in the future. We define G_t as the total discounted return from the current time step t as

$$G_t = r_{t+1} + \gamma r_{t+2} + \gamma^2 r_{t+3} + \cdots = \sum_{k=0}^{\infty} \gamma^k r_{t+k+1}, \tag{7.17}$$

where $\gamma \in [0, 1]$ is the discount factor, determining the relative importance of immediate rewards compared to future rewards. In particular, if the discount factor γ tends to 0, the agent cares only about the immediate rewards, while on the other hand, if γ tends to 1, the agent gives equal importance to immediate rewards and future rewards. A γ between 0 and 1 balances the trade-off between short-term and long-term rewards.

An agent's behavior is defined by a policy π, mapping states $\mathbf{s}$ to a probability distribution over the actions $\mathbf{a}$, which can be written as

$$\pi\left(\mathbf{a}|\mathbf{s}\right) = \mathcal{P}\left(\mathbf{a} = \mathbf{a}_t \,|\mathbf{s} = \mathbf{s}_t\right), \tag{7.18}$$

where it indicates that if an agent follows a policy π at time step t and the state $\mathbf{s}_t$, then the policy defines the probability of an action $\mathbf{a}_t$.
The action-value function $Q_\pi(\mathbf{s}, \mathbf{a})$ is the expected long-term return, from the state $\mathbf{s}$, taking the action $\mathbf{a}$, and following the policy π, which can be expressed as

$$Q_\pi(\mathbf{s}, \mathbf{a}) = \mathbb{E}_\pi\left[G_t|\mathbf{s} = \mathbf{s}_t, \mathbf{a} = \mathbf{a}_t\right]. \tag{7.19}$$

In RL, the Bellman equation is used to define a recursive relationship between Q-values [4]. Specifically, based on the Bellman equation, the Q-value of a state-action pair can be expressed into an immediate reward plus a discounted Q-value of the subsequent state as

$$Q_\pi\left(\mathbf{s}_t, \mathbf{a}_t\right) = \mathbb{E}_\pi\left[r_{t+1} + \gamma Q_\pi\left(\mathbf{s}_{t+1}, \mathbf{a}_{t+1}\right)\right], \tag{7.20}$$

where (7.20) is the foundation equation to update Q-value for many RL algorithms.

Relation to DDPG

In DDPG, we use neural networks as function approximators to approximate both Q-value (i.e., critic network) and policy network (i.e., actor network). Considering $\mathbf{W}^Q$ and $\mathbf{W}^\mu$ to denote critic and actor network, respectively, then, we can optimize the weight by minimizing the MSE loss $\mathcal{L}$ as

$$\mathcal{L}_{\text{MSE}} = \left(y_t - Q\left(\mathbf{s}_t, \mathbf{a}_t | \mathbf{W}^Q\right)\right)^2, \tag{7.21}$$

where y_t can be written as

$$y_t = r_{t+1} + \gamma Q\left(\mathbf{s}_{t+1}, \mu\left(\mathbf{s}_{t+1} | \mathbf{W}^\mu\right)\right). \tag{7.22}$$

It can be seen that (7.21) and (7.22) are similar to (7.6) and (7.7), respectively. Specifically, (7.6), and use the same approach, only with the consideration of having replay buffer and target actor and critic networks.

References

1. N.C. Luong et al., Applications of deep reinforcement learning in communications and networking: a survey. IEEE Commun. Surv. Tuts. **21**(4), 3133–3174 (2019)
2. A. Feriani, E. Hossain, Single and multi-agent deep reinforcement learning for AI-enabled wireless networks: a tutorial. IEEE Commun. Surv. Tuts. **23**(2), 1226–1252 (2021)
3. P.S. Bithas, E.T. Michailidis, N. Nomikos, D. Vouyioukas, A.G. Kanatas, A survey on machine-learning techniques for UAV-based communications. Sensors (Basel, Switzerland) **19** (2019). https://api.semanticscholar.org/CorpusID:208357295
4. T.P. Lillicrap et al., Continuous control with deep reinforcement learning (2015). CoRR, vol. abs/1509.02971. https://api.semanticscholar.org/CorpusID:16326763
5. A. Hajijamali Arani, M.M. Azari, P. Hu, Y. Zhu, H. Yanikomeroglu, S. Safavi-Naeini, Reinforcement learning for energy-efficient trajectory design of UAVs. IEEE Internet Things J. **9**(11), 9060–9070 (2022)
6. H. Dong, D. Zhao, D. Huang, K. Yan, W. Ren, SARSA λ reinforcement learning based path planning of unmanned aerial vehicles, in *Proceedings of 2024 IEEE 13th Data-Driven Control Learning System Conference (DDCLS)* (2024), pp. 1099–1105
7. M. Muzammul, M. Assam, Y.Y. Ghadi, N. Innab, M. Alajmi, T.J. Alahmadi, IR-QLA: machine learning-based Q-learning algorithm optimization for UAVs faster trajectory planning by instructed- reinforcement learning. IEEE Access **12**, 91300–91315 (2024)
8. I. Zaghbani, R. Jarray, S. Bouallègue, Comparative study of Q-learning and SARSA algorithms for UAV path planning in 3D environments, in *Proceedings of 2024 IEEE 28th International Conference on Intelligent Engineering Systems (INES)* (2024), pp. 000245–000250
9. N.Q.H. Othman, J. Li, Q. Yang, Multi-agent reinforcement learning aided resource allocation with SARSA in UAV networks, in *Proceedings of 2023 IEEE International Conference on Signal Processing, Communications, and Computing (ICSPCC)* (2023), pp. 1–6
10. H. Huang, Y. Yang, H. Wang, Z. Ding, H. Sari, F. Adachi, Deep reinforcement learning for UAV navigation through massive MIMO technique. IEEE Trans. Veh. Technol. **69**(1), 1117–1121 (2020)
11. S.P. Gopi, M. Magarini, Reinforcement learning aided UAV base station location optimization for rate maximization. Electronics **10**(23), (2021). ISSN: 2079-9292. https://www.mdpi.com/2079-9292/10/23/2953
12. C. Wang, J. Wang, J. Wang, X. Zhang, Deep-reinforcement-learning-based autonomous UAV navigation with sparse rewards. IEEE Internet Things J. **7**(7), 6180–6190 (2020)

13. C. Zhang, Z. Li, C. He, K. Wang, C. Pan, Deep reinforcement learning based trajectory design and resource allocation for UAV-assisted communications. IEEE Commun. Lett. **27**(9), 2398–2402 (2023)
14. H. Wang, H. Zhang, X. Liu, K. Long, A. Nallanathan, Joint UAV placement optimization, resource allocation, computation offloading for THz band: a DRL approach. IEEE Trans. Wireless Commun. **22**(7), 4890–4900 (2023)
15. S. Khairy, P. Balaprakash, L.X. Cai, Y. Cheng, Constrained deep reinforcement learning for energy sustainable multi-UAV based random access IoT networks with NOMA. IEEE J. Sel. Areas Commun. **39**(4), 1101–1115 (2021)
16. M. Ghadaksaz, M. Mahmood, T. Le-Ngoc, UAV-assisted enhanced coverage and capacity in dynamic MU-mMIMO IoT systems: A deep reinforcement learning approach, in *Proceedings of the 2024 EuCNC & 6G Summit* (2024), pp. 717–722
17. R. Méndez-Rial, C. Rusu, N. González-Prelcic, A. Alkhateeb, R.W. Heath, Hybrid MIMO architectures for millimeter wave communications: phase shifters or switches? IEEE Access **4**, 247–267 (2016)
18. A. Koc, A. Masmoudi, T. Le-Ngoc, 3D angular-based hybrid precoding and user grouping for uniform rectangular arrays in massive MU-MIMO systems. IEEE Access **8**, 84689–84712 (2020)

Chapter 8
A Codebook-Based DRL Approach for UAV Deployment and Trajectory Design

In Chap. 7, we used DDPG as an RL approach, where it showed a performance close to the NI, PSO-based solution, while reducing the runtime by 31.5% in dynamic environments, by taking advantage of transfer learning. Although the runtime reduction is substantial, this might not be sufficient in dynamic scenarios, where users are constantly on the move, and the UAV needs to change its location frequently. This brings us to introduce a novel low-complexity DDPG codebook-based approach for the UAV deployment problem.

Codebook-based solutions have been investigated in the UAV-assisted communications to reduce the computational complexity [1, 2]. For instance, the authors in [1] propose a joint codebook design and beam training for mmWave-aided UAV communications. Specifically, the authors first design a 3-D codebook that provides flexible access for Internet of Things (IoT) devices and achieve the optimal system throughput. Then, based on the designed codebook, an angle for angle forecast-based fast beam alignment (AFFBA) mechanism is proposed, which predicts the range of ideal AoD from the current UAV's beam. Their results show that this approach significantly reduces the beam sweeping space and improves the spectral efficiency compared to exhaustive training. In [2], the authors propose a low-complexity codebook-based HBF for mmWave-UAV communications. Given a pretrained analog codebook, the authors consider selecting the beams that could be more correlated to the space of the channel matrix. Their results show that the proposed low-complexity method can be applicable to the UAV-based system with many users.

In this chapter, we consider a UAV as a DF relay between several remote, moving users and a static BS, where the only means of communication between them is the UAV relay. Considering HBF for both BS and UAV, a novel low-complexity DDPG-C-UD is proposed, consisting of an offline agent training phase, and an online UAV deployment prediction to achieve maximum AR. Subsequently, we extend the proposed DDPG-C-UD to multiple UAV scenarios. The illustrative results show that the DDPG-C-UD can attain a performance close to the DDPG-based solution

T. Le-Ngoc et al., *AI-Enabled UAV-Assisted Massive MIMO*, Wireless Networks,
https://doi.org/10.1007/978-3-032-07872-8_8

in both static and dynamic environments, while reducing the runtime by 99%, making the DDPG codebook-based solution a promising implementation for real-time online applications. Deploying multiple UAVs can enhance the total AR by 421.9%, significantly enhancing the system's capacity to support more users/devices [3].

The rest of this chapter is organized as follows. Section 8.1 introduces the system model and problem formulation followed by a detailed explanation of the DDPG codebook-based solution in Sect. 8.2. Subsequently, Sect. 8.3 presents the illustrative result for the proposed algorithmic solutions. Finally, Sect. 8.4 provides the chapter's conclusion.

8.1 System Model and Problem Formulation

We consider the same system proposed in Sect. 7.2. The BS is equipped with N_T antennas, and the UAV relay has N_r antennas for reception and N_t antennas for transmission to K users. Both the BS and UAV employ HBF, with the BB and RF stages at the BS represented by $\mathbf{B}_b \in \mathbb{C}^{N_{RF_b} \times K}$ and $\mathbf{F}_b \in \mathbb{C}^{N_T \times N_{RF_b}}$, where N_{RF_b} is the number of RF chains. Similarly, the UAV's receive stages are given by $\mathbf{B}_{u,r} \in \mathbb{C}^{K \times N_{RF_u}}$ and $\mathbf{F}_{u,r} \in \mathbb{C}^{N_{RF_u} \times N_r}$, while its transmit stages are represented by $\mathbf{B}_{u,t} = [\mathbf{b}_{u,t,1}, \cdots, \mathbf{b}_{u,t,K}] \in \mathbb{C}^{K \times N_t}$ and $\mathbf{F}_{u,t} = [\mathbf{f}_{u,t,1} \cdots, \mathbf{f}_{u,t,N_{RF_u}}] \in \mathbb{C}^{N_t \times N_{RF_u}}$. The channel between the BS and UAV is modeled as $\mathbf{H}_1 \in \mathbb{C}^{N_r \times N_T}$, and the channel between the UAV and users as $\mathbf{H}_2 \in \mathbb{C}^{K \times N_t}$. Additionally, we assume the environmental noise follows a distribution $\mathcal{CN}(\mathbf{0}, \sigma_n^2)$. Then, by defining $\mathbf{Q}_1^{-1} = (\sigma_n^2 \mathbf{B}_{u,r}\mathbf{F}_{u,r})^{-1}\mathbf{F}_{u,r}^H\mathbf{B}_{u,r}^H$ and $\mathcal{H}_1 = \mathbf{F}_{u,r}\mathbf{H}_1\mathbf{F}_b$, the AR for the first link can be written as

$$\mathrm{R}_1(\mathbf{F}_b, \mathbf{B}_b, \mathbf{F}_{u,r}, \mathbf{B}_{u,r}) = \log_2 |\mathbf{I}_K + \mathbf{Q}_1^{-1}\mathbf{B}_{u,r}\mathcal{H}_1\mathbf{B}_b\mathbf{B}_b^H\mathcal{H}_1^H\mathbf{B}_{u,r}^H|. \tag{8.1}$$

Subsequently, the AR for the second link can be expressed as

$$\mathrm{R}_2(\mathbf{F}_{u,t}, \mathbf{B}_{u,t}, x_u, y_u) = \mathbb{E}\left\{\sum_{k=1}^{K} \mathbb{E}\left[\log_2(1 + \mathrm{SINR}_k)\right]\right\}, \tag{8.2}$$

where $[x_u, y_u]$ is the UAV's 2-D location and SINR_k is given as

$$\mathrm{SINR}_k = \frac{|\mathbf{h}_{2,k}^H\mathbf{F}_{u,t}\mathbf{b}_{u,t,k}|^2}{\sum_{\hat{k}\neq k}^{K} |\mathbf{h}_{2,\hat{k}}^H\mathbf{F}_{u,t}\mathbf{b}_{u,t,\hat{k}}|^2 + \sigma_n^2}. \tag{8.3}$$

Here, $\mathbf{h}_{2,g} \in \mathbb{C}^{N_t}$ is the channel vector between UAV and kth user. We account for fading and shadowing effects by considering a mmWave channel model, where the details can be found in Sect. 4.2.2.

8.1.1 Problem Formulation

Given the BB stages $\mathbf{B}_b$, $\mathbf{B}_{u,r}$, and $\mathbf{B}_{u,t}$, along with the RF stages $\mathbf{F}_b$, $\mathbf{F}_{u,r}$, and $\mathbf{F}_{u,t}$, as developed in Sects. 4.3.2 and 4.3.1, the UAV deployment problem for maximizing the AR can be formulated as follows:

$$\begin{aligned} \max_{\{\mathbf{x}_u\}} \quad & \mathrm{R_T}\left(\mathbf{F}_b, \mathbf{B}_b, \mathbf{F}_{u,r}, \mathbf{B}_{u,r}, \mathbf{F}_{u,t}, \mathbf{B}_{u,t}, \mathbf{x}_u\right) \\ \text{s.t.} \quad & C_1 : x_{\min} \leq x_u \leq x_{\max}, \quad y_{\min} \leq y_u \leq y_{\max}, \end{aligned} \tag{8.4}$$

where $\mathrm{R_T} = \frac{1}{2}\min(\mathrm{R}_1, \mathrm{R}_2)$ is the total transmission rate from BS to users, $\mathbf{x}_u = [x_u, y_u]$ is the 2-D deployment within a given flying span, and $[x_{\min}, x_{\max}]$ and $[y_{\min}, y_{\max}]$ represent the UAV deployment range in the x-axis and y-axis, respectively. To overcome the problem stated in (8.4), we propose a novel low-complexity DDPG codebook-based solution for the UAV deployment problem in the next section.

8.2 Low-Complexity DDPG-C-UD

In Sect. 7.3, we proposed a DDPG-based solution to address the UAV deployment problem, where it achieved near-optimal performance in solving the UAV deployment problem in both dynamic and static environments. However, in real-world scenarios where user locations are constantly changing, the proposed DDPG-based solution may struggle to adapt in a timely manner, making it impractical for real-time online applications in UAV-assisted MU-mMIMO communications systems. To address this, we propose a low-complexity DDPG codebook-based algorithm that can achieve near-optimal AR with much-reduced complexity (i.e., runtime). The proposed algorithm has two phases, as depicted in Fig. 8.1: (1) offline agent training and (2) online UAV deployment. These phases are discussed in detail in the following subsections.

8.2.1 Offline Agent Training (Phase 1)

This phase involves training the agent (i.e., optimizing the weights of the (target) actor and (target) critic networks) using the DDPG algorithm for a total of M_e different environments. Here, each environment is designed to represent various user distributions, with the selection carefully made to ensure that the final deployments—referred to as UAV stations—are distributed in a way that evenly covers the entire deployment area, thereby preventing any regions from being left unserved. The UAV's final 2-D deployment in the mth environment is denoted by

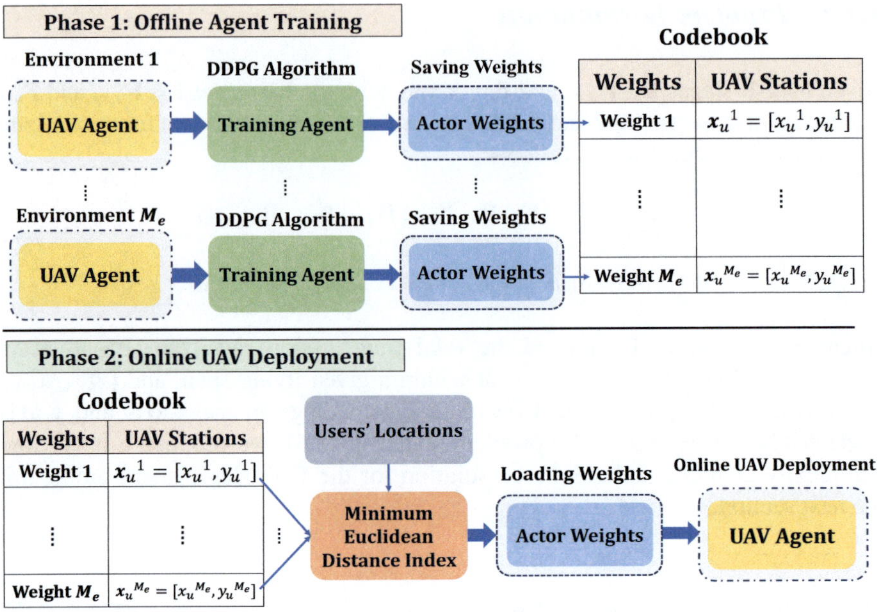

Fig. 8.1 Block diagram of offline agent training and online UAV deployment in DDPG codebook-based algorithms

$\mathbf{x}_u^m = \left[x_u^m, y_u^m\right]^{\mathrm{T}}$, where $m = \{M_1, \cdots, M_e\}$. Here, each environment has its own set of actor and critic networks, represented by $\mu(\mathbf{s}|\mathbf{W}_m^{\mu})$ and $Q(\mathbf{s}, \mathbf{a}|\mathbf{W}_m^{Q})$, where $\mathbf{W}_m^{\mu}$ and $\mathbf{W}_m^{Q}$ are the weights of the critic and actor networks in the mth environment, respectively. Similarly, the corresponding target networks are denoted as $\mu'(\mathbf{s}|\mathbf{W'}_m^{\mu})$ for the target actor and $Q'(\mathbf{s}, \mathbf{a}|\mathbf{W'}_m^{Q})$ for the target critic, where $\mathbf{W}_m^{\mu'}$ and $\mathbf{W}_m^{Q'}$ represent the weights of the target networks. Subsequently, the weights of the actor network $\mathbf{W}_m^{\mu}$, along with the UAV's final 2-D deployment $\mathbf{x}_u^m$ in each environment, are stored in a codebook with M_e rows. The actor weights are stored in the codebook because, as a policy-based RL algorithm, DDPG requires only the actor network during online deployment. While the critic network plays a crucial role during training, it is not required during the online phase, where the actor alone determines the actions to be taken.

To successfully implement the DDPG algorithm in this training phase, it is crucial to define the agent's states ($\mathbf{s}$), actions ($\mathbf{a}$), and rewards (r) effectively. Furthermore, the architecture of the actor and critic networks plays a significant role in determining the algorithm's performance. Detailed information on the selection of states, actions, reward functions, and the design of the actor and critic networks can be found in Sect. 7.3.2. The summary of the offline training phase is provided in Algorithm 8.1.

Algorithm 8.1: DDPG-C-UD: offline agent training

for m=*1*:M_e **do**
  Establish environment E_m.
  Randomly initialize actor $\mu(\mathbf{s}|\mathbf{W}_m^{\mu})$ and critic $Q(\mathbf{s}, \mathbf{a}|\mathbf{W}_m^{Q})$ networks with weights $\mathbf{W}_m^{\mu}$ and $\mathbf{W}_m^{Q}$ corresponding to E_m.
  Initialize target network Q' and μ' with weights $\mathbf{W}_m^{Q'} \leftarrow \mathbf{W}_m^{Q}$, $\mathbf{W}_m^{\mu'} \leftarrow \mathbf{W}_m^{\mu}$ corresponding to E_m.
  Initialize replay buffer D.
  for *episode*=*1*:M **do**
    Receive initial observation state $\mathbf{s}_1$.
    for t=*1*:T **do**
      Select $\mathbf{a}_t = \mu(\mathbf{s}_t|\mathbf{W}_m^{\mu}) + \mathcal{N}$.
      Execute action $\mathbf{a}_t$ and observe reward r_t and new state $\mathbf{s}_{t+1}$.
      Store transition $(\mathbf{s}_t, \mathbf{a}_t, r_t, \mathbf{s}_{t+1})$ in D.
      Sample a random minibatch of B_S transitions $(\mathbf{s}_i, \mathbf{a}_i, r_i, \mathbf{s}_{i+1})$ from D.
      Set $y_i = r_i + \gamma Q'(\mathbf{s}_{i+1}, \mu'(\mathbf{s}_{i+1}|\mathbf{W}_m^{\mu'})|\mathbf{W}_m^{Q'})$.
      Update the critic by minimizing $\mathcal{L}_{\text{MSE}} = \frac{1}{B_S}\sum_i (y_i - Q(\mathbf{s}_i, \mathbf{a}_i|\mathbf{W}_m^{Q}))^2$.
      Update the actor using policy gradient:
      $\nabla_{\mathbf{W}_m^{\mu}} J \approx \frac{1}{B_S}\sum_i \nabla_a Q(\mathbf{s}, \mathbf{a}|\mathbf{W}_m^{Q})|_{\mathbf{s}=\mathbf{s}_i, a=\mu_{\mathbf{s}_i}} \nabla_{\mathbf{W}_m^{\mu}} \mu(\mathbf{s}|\mathbf{W}_m^{\mu})|_{\mathbf{s}_i}$.
      Update the target networks via:
      $\mathbf{W}_m^{Q'} \leftarrow \nu \mathbf{W}_m^{Q} + (1-\nu)\mathbf{W}_m^{Q'}$,
      $\mathbf{W}_m^{\mu'} \leftarrow \nu \mathbf{W}_m^{\mu} + (1-\nu)\mathbf{W}_m^{\mu'}$.
    end
  end
  Store the weights $\mathbf{W}_m^{\mu}$.
end

8.2.2 *Online UAV Deployment (Phase 2)*

In phase 2, the algorithm determines the minimum distance between the users' locations and the UAV stations, based on the following selection criteria:

$$m_{\min} = \underset{m \in \{1,\ldots,M_e\}}{\arg\min} \sum_{k=1}^{K} \sqrt{(x_k - x_u^m)^2 + (y_k - y_u^m)^2}, \tag{8.5}$$

where $m_{\min}$ denotes the index of the closest station. It is noteworthy that since the UAV operates at a constant altitude, only the 2-D distance between the users and the UAV is sufficient to determine the closest station, which can reduce computational complexity. Subsequently, the actor weights corresponding to the $m_{\min}$ station are first extracted and then loaded to the UAV agent as follows:

$$\mathbf{W}^{\mu} \leftarrow \mathbf{W}_{m_{\min}}^{\mu}. \tag{8.6}$$

Algorithm 8.2: DDPG-C-UD: online UAV deployment

Randomly initialize actor $\mu(\mathbf{s}|\mathbf{W}^{\mu})$ and critic $Q(\mathbf{s}, \mathbf{a}|\mathbf{W}^{Q})$ networks with weights $\mathbf{W}^{\mu}$ and $\mathbf{W}^{Q}$.
Initialize target network Q' and μ' with weights $\mathbf{W}^{Q'} \leftarrow \mathbf{W}^{Q}$, $\mathbf{W}^{\mu'} \leftarrow \mathbf{W}^{\mu}$.
Find the closest UAV station to the users using (8.5).
Load the actor weights corresponding to the closest station to the UAV agent using (8.6).
Receive initial observation state $\mathbf{s}_1$.
for t=*1*:T **do**
    Select action $\mathbf{a}_t$ using (8.7).
    Execute action $\mathbf{a}_t$ and observe new state $\mathbf{s}_{t+1}$.
end

Consequently, given a state $\mathbf{s}_t$, the UAV will act as follows:

$$\mathbf{a}_t = \mu(\mathbf{s}_t|\mathbf{W}^{\mu}_{m_{\min}}), \tag{8.7}$$

where the exploration noise is not required during the online deployment of the UAV. The summary of phase 2 can be found in Algorithm 8.2.

The codebook size (i.e., the number of stations) significantly influences the performance of the DDPG codebook-based solution. Increasing the number of stations improves the precision of UAV positioning at the cost of longer training times during the offline phase. Additionally, the selection of environments should ensure that UAV stations evenly cover the entire area, preventing any regions from being left unserved. In this chapter, we consider three different DDPG codebook-based solutions with different codebook sizes: (1) 1×1 DDPG-C-UD, (2) 3×3 DDPG-C-UD, and (3) 5×5 DDPG-C-UD. For each of these codebook-based solutions, we divide the area equally to avoid any uncovered regions, as illustrated in Fig. 8.2.

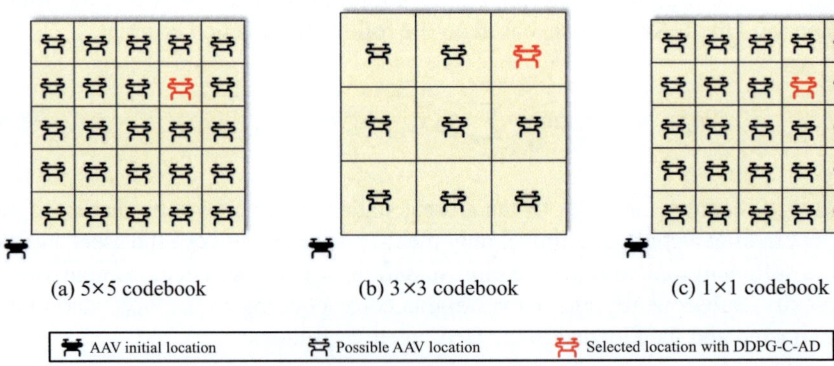

Fig. 8.2 UAV stations for: (**a**) 5×5 codebook, (**b**) 3×3 codebook, and (**c**) 1×1 codebook

8.3 Illustrative Results

In this section, we first provide illustrative results on the performance of DDPG-C-UD in comparison with other schemes. Then, we present the trajectory design for single and multiple UAV scenarios. Table 8.1 outlines the simulation setup based on the 3-D microcell scenario [4] for the results discussed hereafter.

8.3.1 DDPG-C-UD Performance

In the first place, we assume that users are positioned at fixed locations and examine two scenarios: (1) a narrow-range user distribution and (2) a wide-range user distribution. For the narrow-range scenario, the BS is positioned at $(x_b, y_b, z_b) = (0, 0, 10)$, while the UAV begins at $(x_u, y_u, z_u) = (50, 50, 20)$. In this case, $K = 4$ users are randomly placed at a significant distance from the BS, specifically within $(x_k, y_k) \in [90, 100]$. In the wide-range scenario, while the UAV's initial position and the BS location remain the same, the users are randomly distributed over a broader area within $(x_k, y_k) \in [50, 100]$.

Using the proposed DDPG hyper-parameters and architecture discussed in Sect. 7.4.1, we create the aforementioned codebooks using narrow-range user distribution environments. For instance, to establish a station at $(x_u, y_u) = (75, 75, 20)$, we trained the agent in an environment where users are located within $(x_k, y_k) \in [70, 80]$ and then stored the networks' weights. Then, after finishing the offline phase (i.e., phase 1) of the DDPG codebook-based approach, we compare the performances of DDPG-based and DDPG codebook-based solutions.

Table 8.1 Simulation parameters

Number of antennas		$(N_T, N_t, N_r) = 144$	
BS height	UAV height	10 m	20 m
UAV x-axis range	UAV y-axis range	$[x_{\min}, x_{\max}] = [0, 100]$ m	$[y_{\min}, y_{\max}] = [0, 100]$ m
UAV x-axis movement	UAV y-axis movement	$[a_{x,\min}, a_{x,\max}] = [-1, 1]$ m	$[a_{y,\min}, a_{y,\max}] = [-1, 1]$ m
# of paths	Path loss exponent	$L = 10$	3.6
Noise PSD	Reference path loss α	− 174 dBm/Hz	61.34 dB
Frequency	Channel bandwidth	28 GHz	100 MHz
Mean AAoD/AAoA (1st link)	Mean AAoD (2nd link)	120°	$\phi_g = 21° + 120°(g - 1)$
Mean EAoD/EAoA (1st link)	Mean EAoD (2nd link)	60°	$\theta_g = 60°$
Azimuth/elevation angle spread	# of network realization	±10°	2000

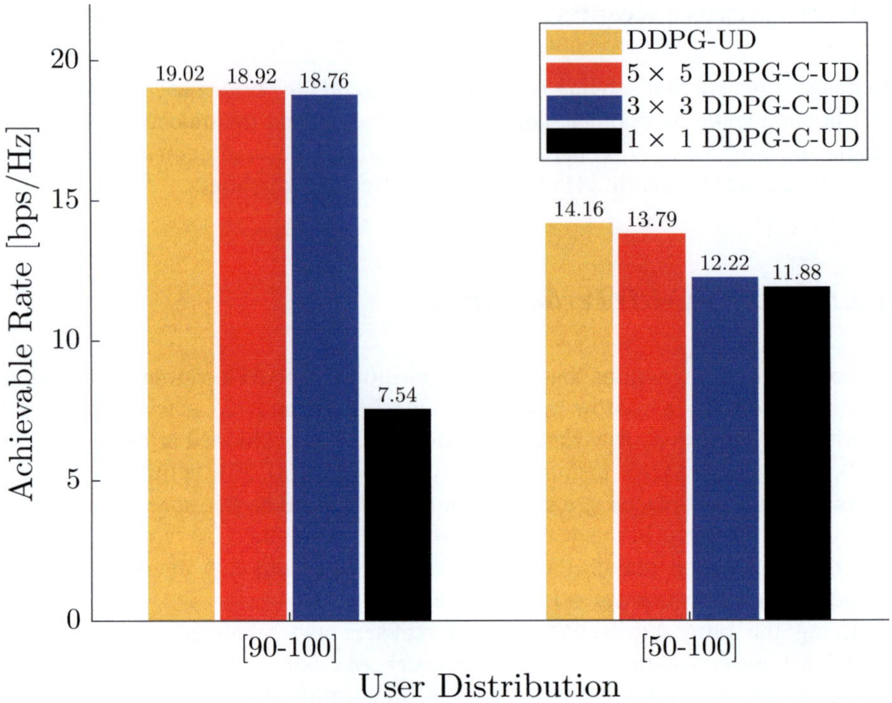

Fig. 8.3 Total AR for different user distributions at $P_T = 30$ dBm

Figure 8.3 demonstrates the AR for DDPG-based and DDPG codebook-based algorithms. It can be seen that 5×5 DDPG-C-UD can achieve 99.47% of DDPG-UD AR when user distribution is [90-100], while with a small degradation, 3×3 DDPG-C-UD accomplishes 98.63% of DDPG-UD performance. However, 1×1 DDPG-C-UD shows a poor performance by only achieving 39.64% of the benchmark AR. 5×5 DDPG-C-UD also shows a close performance to DDPG-UD when user distribution is [50-100] by achieving 97.39% of the DDPG-UD AR, while on the other hand, 3×3 DDPG-C-UD and 1×1 DDPG-C-UD reach 86.30% and 83.90% of the benchmark, respectively. These numerical results reveal that 5×5 DDPG-C-UD accomplishes a near-optimal AR for both wide-range and narrow-range distribution. Moreover, although 3×3 DDPG-C-UD achieves a close performance in [90-100] user distribution, it has an 11.09% of degradation in AR with respect to 5×5 DDPG-C-UD in [50-100] one. Finally, it can be seen that 1×1 DDPG-C-UD has the worst performance in either of the scenarios, specifically when users are far from the center, i.e., [90-100] user distribution.

After discussing the static environment, we explore a more realistic scenario where users can change their locations during training, reflecting dynamic, real-world environments. We assume the BS is positioned at $(x_b, y_b, z_b) = [0, 0, 10]$. The users' locations are then randomly assigned across six different distributions

$l \in \{l_1, l_2, \ldots, l_6\}$ as follows:

$$(x_{k,l}, y_{k,l}) = \begin{cases} x_{k,l} \in [50, 60], \ y_{k,l} \in [50, 60] \text{ for } l_1 \\ x_{k,l} \in [60, 70], \ y_{k,l} \in [60, 70] \text{ for } l_2 \\ x_{k,l} \in [70, 80], \ y_{k,l} \in [70, 80] \text{ for } l_3 \\ x_{k,l} \in [70, 80], \ y_{k,l} \in [80, 90] \text{ for } l_4 \\ x_{k,l} \in [70, 80], \ y_{k,l} \in [90, 100] \text{ for } l_5 \\ x_{k,l} \in [80, 90], \ y_{k,l} \in [90, 100] \text{ for } l_6 \end{cases} \tag{8.8}$$

where $x_{k,l}$ and $y_{k,l}$ denote the kth user x-axis and y-axis ranges during lth distribution, respectively. Furthermore, UAV starts its initial location at $(x_u, y_u, z_u) = (50, 50, 20)$ and finds the optimal deployment $\mathbf{x}_u^{(1)} = \left[x_u^{(1)}, y_u^{(1)}\right]^{\mathrm{T}}$ for l_1, which is now used as its initial location for the next user distribution (i.e., l_2).

Figure 8.4 compares the AR forDDPG-based solution and DDPG codebook-based solutions for $K = 4$ at $P_T = 30$ dBm, which shows that 5×5 DDPG-C-UD can achieve a performance close to DDPG-UD. For instance, 5×5 DDPG-C-UD accomplishes 99.94% of DDPG-UD for l_2, while 3×3 DDPG-C-UD and 1 × 1 DDPG-C-UD achieve 98.87% and 69.17% as compared to DDPG-UD,

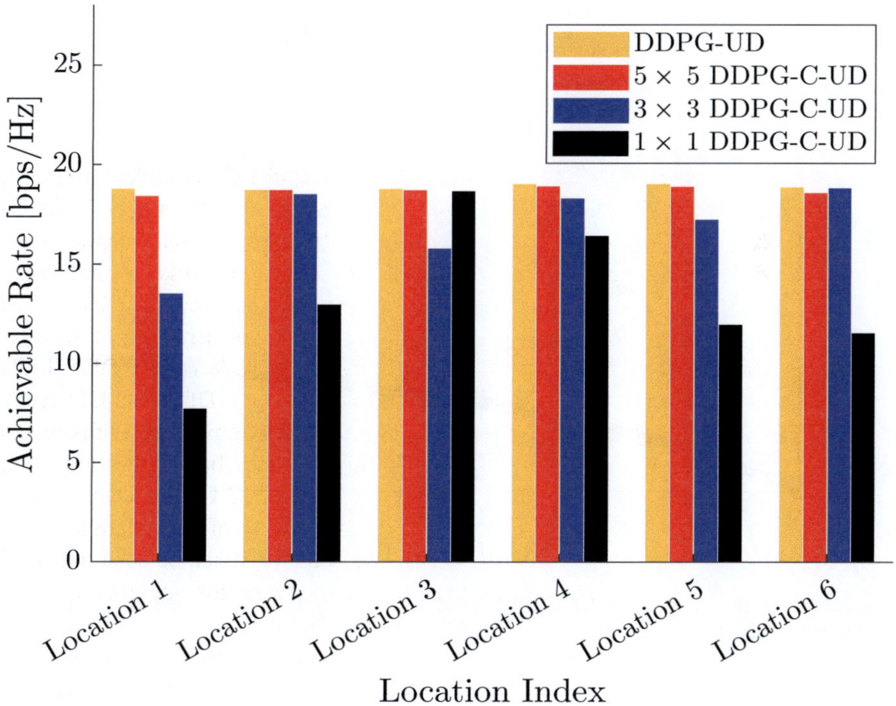

Fig. 8.4 Total AR for different locations at $P_T = 30$ dBm

respectively. Also, it can be seen that 5×5 DDPG-C-UD provides 18.87 bps/Hz and achieves 99.32% of DDPG-UD solution. With some degradation, 3×3 DDPG-C-UD achieves 90.60%, and 1×1 DDPG-C-UD reaches only 62.72% of DDPG-UD performance for l_6. Moreover, it is noticeable that at l_3 the 1×1 DDPG-C-UD performs better than the 3×3 DDPG-C-UD, and at l_6, the 3×3 DDPG-C-UD performs better than the 5×5 DDPG-C-UD. The better performance of less complex codebooks occurs because the user distribution is sometimes very close to one of their stations, causing a performance spike. On average, 5×5 DDPG-C-UD achieves 99.16% of DDPG-UD achieved rate over these locations. As expected, and with performance degradation, 3×3 DDPG-C-UD and 1×1 DDPG-C-UD accomplish 90.26% and 69.98% of DDPG-UD performance, respectively.

In Fig. 8.5, we present the runtime for DDPG-based and DDPG codebook-based solutions for a different number of user locations. We provide the runtime results for 100 network realizations for both DDPG-based and DDPG codebook-based algorithms. It is worthwhile to note that both DDPG-based and DDPG

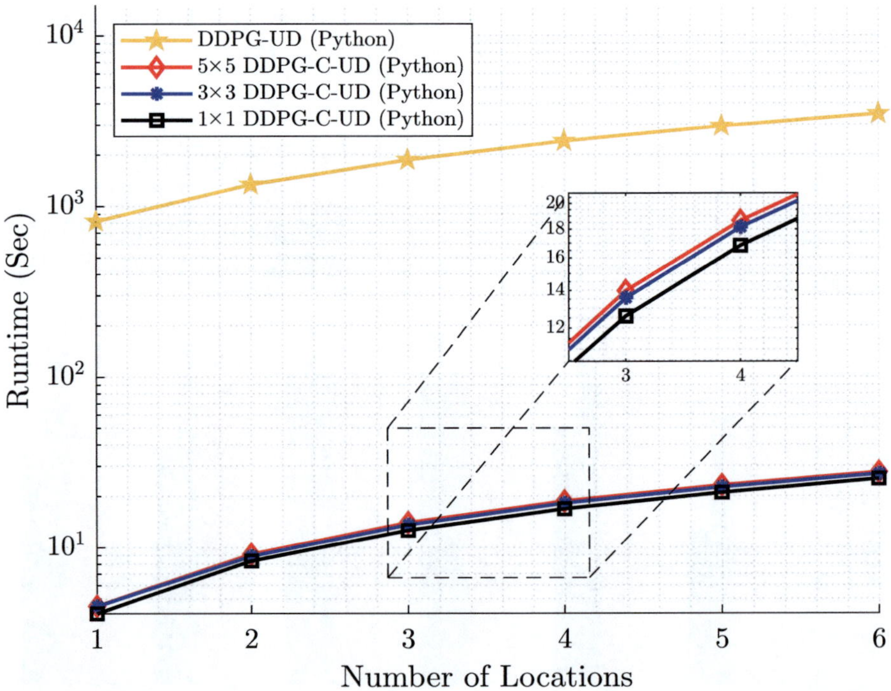

Fig. 8.5 Runtime comparison for DDPG-based and DDPG codebook-based solutions for a different number of user locations

codebook-based algorithms are run on Python.[1] We observe that the proposed DDPG codebook-based solutions outperform the computational complex DDPG-based algorithm by significantly reducing the runtime. For instance, with three user locations, the DDPG-UD requires 1860.8 sec to find the optimal locations, whereas only 13.973 sec is enough for 5×5 DDPG-C-UD to predict UAV location in phase 2. Therefore, 5×5 DDPG-C-UD needs only about 0.75% of DDPG-UD without impacting the total AR performance. Also, it can be seen that if we increase the number of user locations to 6, DDPG-UD can take around 3467.6 sec, while 5×5 DDPG-C-UD requires only 27.528 sec, which is almost 0.75% of the DDPG-UD. This means that using 5×5 DDPG-C-UD, we can decrease the runtime by almost 99%, while providing a performance close to DDPG-UD. Moreover, it is noteworthy that although 5×5 DDPG-C-UD has a longer offline phase as compared to 3×3 DDPG-C-UD and 1×1 DDPG-C-UD, in the online phase, the difference is minor where for instance for six user locations, they have 97.80% and 91.68% of 5×5 DDPG-C-UD runtime, respectively.

Next, we analyze and compare the computational complexity of the DDPG-based and DDPG codebook-based solutions. We begin with the computational requirements for the angular-based HBF design. The RF beamformers $\mathbf{F}_b$, $\mathbf{F}_{u,r}$, and $\mathbf{F}_{u,t}$ have complexities of $O(N_T)$, $O(N_r)$, and $O(N_t)$ operations, respectively. For the BB stages, $\mathbf{B}_b$ and $\mathbf{B}_{u,r}$ require $O(N_{RF_b}^3)$ and $O(N_{RF_u}^3)$ operations, respectively. The complexity of $\mathbf{B}_{u,t}$ is $O(N_{RF_u}^3 + KN_{RF_u}^2)$, where $O(KN_{RF_u}^2)$ corresponds to the computation of $\mathcal{H}_2^H\mathcal{H}$, and $O(N_{RF_u}^3)$ is for the matrix inversion. Overall, the HBF designs for BS and UAV require $O(N_T + N_r + N_t + N_{RF_b}^3 + N_{RF_u}^3 + KN_{RF_u}^2)$ operations. Then, we discuss the complexity of the DDPG-UD with the hyper-parameters provided in Table 7.2. For DDPG training, the complexity can be calculated as follows:

- **Forward pass**: ***(a) Actor network***: The actor network consists of two neurons for *input layer*, two *hidden layers* with 20 neurons each, and two neurons for *output layer*. Thus, first we need $O(20 \times 2) = O(40)$ operations to calculate the output of the *input layer*. Consequently, $O(20 \times 20) = O(400)$ and $O(20 \times 2) = O(40)$ operations are required to obtain the output of the first *hidden layer* and second *hidden layer*, respectively. Overall, $O(40 + 400 + 40) = O(480)$ operations are required for the forward pass of actor network. ***(b) Critic network***: The critic network has four neurons for *input layer*, 20 neurons for two *hidden layers*, and one neuron for *output layer*. Similar to the actor network, the complexity of the critic network's forward pass is $O(80 + 400 + 20) = O(500)$. ***(c) Target networks***: The target networks are identical to their primary network architectures. As a result, we require $O(40 + 400 + 40) = O(480)$ and $O(80 + 400 + 20) = O(500)$ for target actor and target critic networks, respectively. Moreover, at each forward pass, we require to update the BB beamforming stages according to

[1] For the Python runtime results, we implement both DDPG-UD and DDPG-C-UD solutions using a PC with Intel Core i7-11800H @ 2.3GHz and 16-GB RAM.

the new observed CSI, with complexity $O\left(N_{RF_b}^3 + N_{RF_u}^3 + KN_{RF_u}^2\right)$.[2] Overall, we require $O\left(1960 + N_{RF_b}^3 + N_{RF_u}^3 + KN_{RF_u}^2\right)$ for the forward pass of DDPG algorithm.

- **Backward pass**: The backward pass step involves computing gradients of the loss function with respect to the weights and biases in both the actor and critic networks. The complexity of the backward pass matches that of the forward pass, as it requires similar operations (without BB stage update). Since the target networks are updated using a soft update rule, we exclude them from the backward pass complexity calculation. Therefore, the total complexity for the backward pass step is $O(500 + 480) = O(980)$ operations.
- **Total complexity**: Considering batch size $B_S = 32$, and training through $T = 100$ time steps with $M = 300$ episodes for serving $K = 4$ users, the total complexity of the DDPG-UD algorithm can be calculated as $O\left(T.M.B_S.3132\right) \approx O\left(3 \times 10^9\right)$.

Compared to the DDPG-based solution, while the DDPG codebook-based approach requires higher computational complexity during the offline training phase, once the codebook is established, it requires significantly less complexity for deployment. For example, in the proposed 5×5 DDPG-C-UD, only $O(25)$ operations are needed to determine the corresponding actor weights, followed by $O(T)$ operations for the deployment phase. This is because the complexity is proportional to the number of layers and neurons in each layer, which is a constant value for a trained actor and critic network. Thus, the 5×5 DDPG-C-UD requires a total of $O\left(T + 25\right) = O\left(125\right)$ operations for UAV deployment, offering a substantial reduction in computational complexity compared to DDPG-UD.

8.3.2 *DDPG Codebook-Based Trajectory Design*

In the following subsections, we demonstrate the UAV path and extend the proposed solution to multiple UAV scenarios.

8.3.2.1 Single UAV

The 5×5 DDPG-C-UD demonstrates strong performance in terms of AR, reaching approximately 99% of the AR achieved by DDPG-UD, as discussed earlier. Next, we illustrate the UAV's flight path under various conditions. Figure 8.6 shows the UAV path in both wide-range and narrow-range user distributions, highlighting how

[2] The RF stages are developed using slow time-varying angular information, and thus, they will not be updated at each time step.

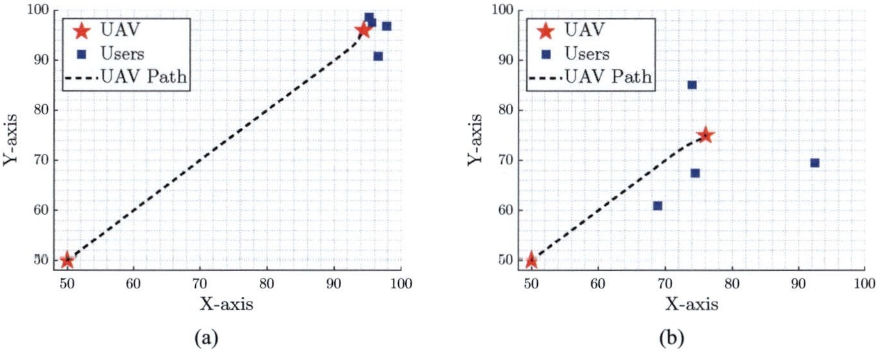

Fig. 8.6 UAV trajectory design for (**a**) narrow-range user distribution and (**b**) wide-range user distribution

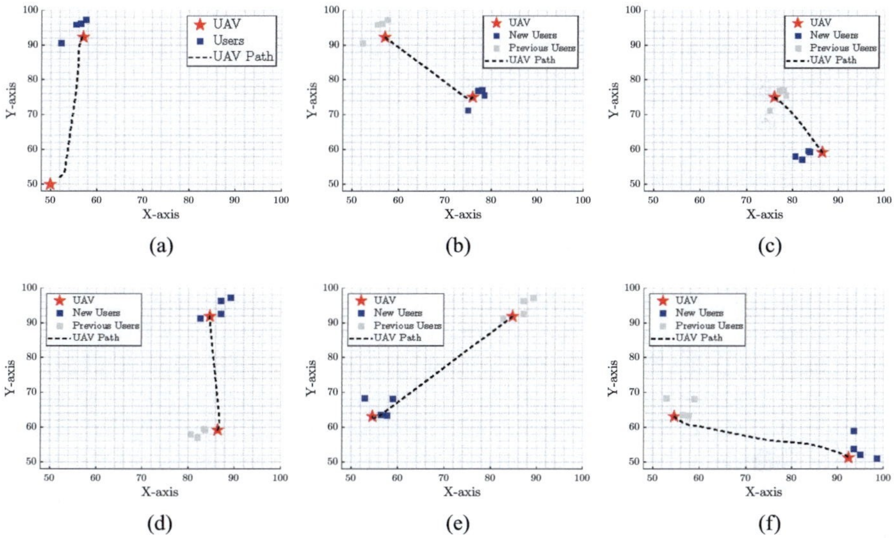

Fig. 8.7 Trajectory design in dynamic environment. (**a**) Trajectory 1. (**b**) Trajectory 2. (**c**) Trajectory 3. (**d**) Trajectory 4. (**e**) Trajectory 5. (**f**) Trajectory 6

the UAV follows a logical route to reach the selected station. We then expand this scenario to dynamic environments, where the UAV needs to frequently change its position to maintain good performance. Figure 8.7 presents different user settings and the corresponding UAV paths. In these cases, the UAV starts from an initial position at $(x_u, y_u, z_u) = [50, 50, 20]$ and moves to the first designated station. The final position of each path becomes the starting point for the next, with the UAV then moving to the next station based on the updated user distribution. Across all paths shown in Fig. 8.7, the UAV follows an almost straight trajectory, demonstrating that

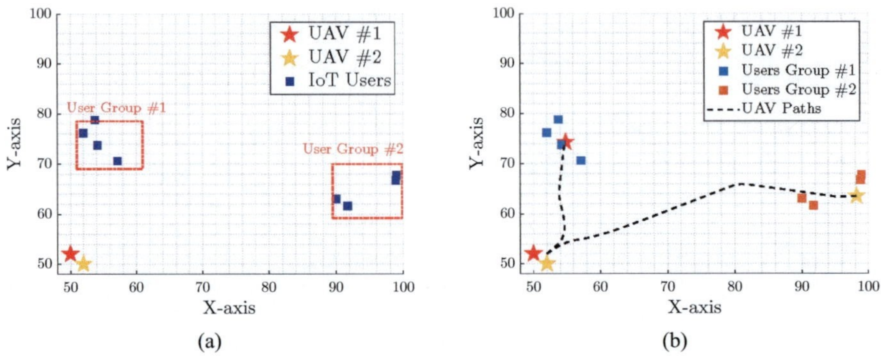

Fig. 8.8 UAV trajectory design for (**a**) narrow-range user distribution and (**b**) wide-range user distribution

the DDPG codebook-based solution is effective not only in achieving a high AR but also in determining efficient flight paths.

8.3.2.2 Multiple UAV

Although a single UAV is sufficient for a limited number of users, for supporting more users or for scenarios where users are far apart, in these situations, more UAVs are required to achieve acceptable AR. Therefore, we incorporate another UAV to serve more users. Figure 8.8 shows a situation where two groups of far-apart users are in the area. In the first place, the users are grouped using the K-means clustering approach. Subsequently, each UAV will be assigned to one of the groups using the online phase of DDPG-C-UD, introduced in Algorithm 8.2. Here, UAVs start their deployment from $(x_u, y_u, z_u) = [50, 50, 20]$.[3] It can be seen in Fig. 8.8 that using 5×5 DDPG-C-UD, UAVs go through a reasonable path to reach the assigned group of users.

Figure 8.9 illustrates the total AR for both single and multiple UAV setups across different transmit powers P_T for the DDPG-C-UD and DDPG-UD solutions. The results clearly show that employing multiple UAVs significantly boosts the total AR compared to using a single UAV, with DDPG-C-UD achieving performance that is very close to that of DDPG-UD. For instance, at $P_T = 30$ dBm, DDPG-C-UD achieves a total AR of 11.05 bps/Hz with a single UAV, whereas with multiple UAV, this value rises dramatically to 37.13 bps/Hz—an increase of 336.02%. In both configurations, DDPG-C-UD achieves 99.55% and 97.67% of the total AR of the DDPG-UD benchmark, respectively. Similarly, at $P_T = 50$ dBm, DDPG-C-UD attains 15.08 bps/Hz in the single UAV case and 63.64 bps/Hz with multiple

[3] For simplicity, we assume there is no collision between UAVs, and the possibility of collision is left as our future work.

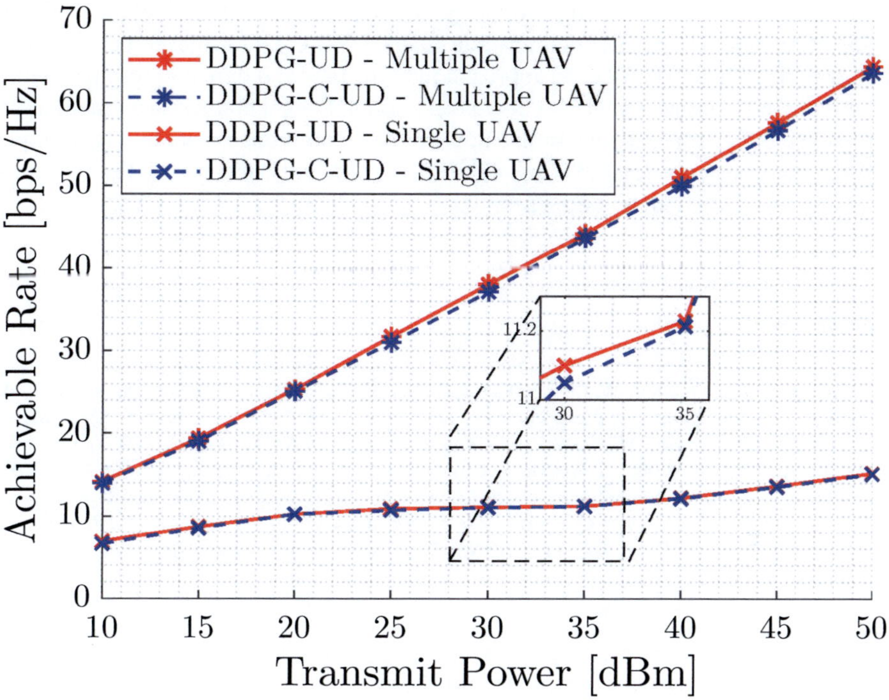

Fig. 8.9 Total AR vs. P_T for single and multiple UAV system

UAVs, resulting in a substantial 421.9% gain in AR. Again, DDPG-C-UD closely matches the DDPG-UD performance, achieving 99.35% and 98.78% of the total AR for single and multiple UAV scenarios, respectively.

The superior performance of multiple UAVs compared to a single UAV can be attributed to two main factors: (a) when users are widely distributed (e.g., as shown in Fig. 8.8a), the UAV in the single configuration often has to position itself farther from users, which increases path loss and consequently reduces AR. (b) As the number of users increases, a single UAV with limited transmit power must allocate smaller portions of power to each user, leading to a decline in the overall AR.

8.4 Concluding Remarks

In this chapter, we investigated a dynamic UAV-assisted communications system where a UAV is deployed as a DF relay between BS and users. We consider that the BS does not have direct access to users due to substantial blockage. Assuming HBF for both BS and UAV, we introduce a novel low-complexity DDPG-C-UD method, including an offline training phase and an online deployment prediction to maximize AR. Then, we expand the proposed DDPG-C-UD to include multiple UAV scenarios. The results demonstrate that DDPG-C-UD achieves performance comparable to the DDPG-based approach in both static and dynamic settings, while also reducing runtime by 99%. This makes the DDPG codebook-based method an appealing choice for real-time online applications. Additionally, deploying multiple UAVs can boost the total AR by 421.9%, significantly increasing the system's capacity to accommodate more users.

References

1. Y. Wang, X. Wen, Y. Chen, W. Jing, Q. Pan, Joint 3d codebook design and beam training for UAV millimeter-wave communications, in *Proceedings of 2019 IEEE 30th International Symposium on Personal, Indoor and Mobile Radio Communications (PIMRC Workshops)* (2019), pp. 1–6
2. Q. Yu, C. Han, J. Wang, L. Bai, Low complexity hybrid beamforming for mmWave-UAV communication systems with a pre-defined codebook, in *Proceedings of the 2nd International Conference on Control and Computer Vision*, ser. ICCCV '19 (Association for Computing Machinery, Jeju, 2019), pp. 109–115. ISBN: 9781450363228. https://doi.org/10.1145/3341016.3341036
3. M. Ghadaksaz, M. Mahmood, T. Le-Ngoc, AAV deployment in IoT networks: a codebook-based reinforcement learning approach. IEEE Internet Things J. **12**(11), 17137–17152 (2025)
4. A. Koc, A. Masmoudi, T. Le-Ngoc, 3D angular-based hybrid precoding and user grouping for uniform rectangular arrays in massive MU-MIMO systems. IEEE Access **8**, 84689–84712 (2020)

Chapter 9
Conclusions and Future Works

In this chapter, we first present a detailed summary of the chapters in this book and then discuss some potential research directions as future works.

9.1 Summary

According to ITU's IMT-2030 (6G) framework, which highlights the need for enhanced capabilities (e.g., peak data rate, latency, connection density, mobility, and reliability) and new capabilities (e.g., ubiquitous coverage, sensing, sustainability, resilience, and positioning), the future wireless networks are expected to adopt aerial architectures to meet these demands. In particular, UAVs serving as relays will play a key role in improving coverage and capacity in future wireless networks. To further support these advancements, mMIMO systems, utilizing large antenna arrays, offer 3-D beamforming capabilities that improve capacity, boost energy efficiency, and reduce interference. However, the FDBF—which requires a large number of power-hungry RF chains—is not practical for UAV-assisted scenarios due to the strict power constraints of UAVs. Furthermore, while the ABF offers lower energy consumption compared to FDBF, it falls short in achieving acceptable performance in MU-mMIMO scenarios due to the phase-only structure. To address this, HBF is proposed as an effective solution. By using fewer RF chains to connect analog RF beamformers with digital BB processing stages, HBF significantly lowers power consumption, while maintaining performance levels comparable to FDBF. Recently, AI-based solutions have emerged as a promising approach to address complex problems in wireless communications and UAV-assisted communications. Within various subfields of AI, we mainly focused on NI- and DL-based algorithms and presented them as proposed solutions for different UAV-assisted communications problems to enhance the coverage and capacity in this book. Specifically, we first studied PSO as an NI algorithm for non-convex UAV-assisted problems to

T. Le-Ngoc et al., *AI-Enabled UAV-Assisted Massive MIMO*, Wireless Networks,
https://doi.org/10.1007/978-3-032-07872-8_9

achieve near-optimal solutions in Part I (i.e., Chaps. 3–4). Then, we presented low-complexity DL-based solutions to reduce the runtime of the proposed PSO-based solutions, while maintaining a satisfactory performance in Part II (i.e., Chaps. 6–8).

Chapter 3 has investigated various UAV-assisted SU-mMIMO systems, in which the UAV is the only means of communication between a BS and a far-located gateway. The HBF has been considered for both BS and gateway combined with different relaying schemes and beamforming setups for the UAV, while PSO-based solution is proposed for UAV location optimization. First, we investigated a UAV acting as an AF relay and having ABF setup. The illustrative results show that the proposed joint ABF- and PSO-based UAV location optimization can achieve higher spectral and energy efficiency compared to existing HBF schemes. Second, we considered the UAV as a DF relay, while having HBF configuration. The numerical results depict that the proposed HBF with PSO-based UAV deployment can significantly improve energy efficiency when compared to FDBF and can find the optimal UAV deployment to enhance coverage and capacity.

In Chap. 4, we have introduced a UAV-assisted terrestrial system to enhance both coverage and capacity in MU-mMIMO systems. Here, the UAV operates as a DF relay to extend the coverage area of BS for obscured users, where the overall capacity is maximized by jointly optimizing the UAV placement, PA to users, and by utilizing angular-based HBF solutions for both BS and UAV. In particular, we develop a PSO-based joint UAV location and PA (PSOLPA) algorithm, which achieved near-optimal sum rate capacity. Secondly, we have introduced a novel buffer-aided transmission to further increase the sum rate capacity. The illustrative results depict that the buffer-aided transmission can enhance the capacity compared to cases without using buffer.

Chapter 4 has extended the UAV-assisted communications to the multiple UAVs-assisted terrestrial networks to support a large number of users, enhancing both coverage and capacity. We have considered multiple UAVs operating as DF relays, which connect the BS to a large number of Internet of Things (IoT) devices and utilized structured sequential optimization to address the multifaceted optimization problem by splitting it into two subproblems. Here, first we have used K-means-based user clustering for UAV-users association. Then, the location of each UAV is optimized jointly with PA using swarm intelligence. The RF stages are designed based on the slow time-varying angular information, while BB stages are designed utilizing the reduced-dimension effective channel matrices. In the results, we have shown that multiple UAV-assisted cooperative relaying systems outperform a single UAV system in practical user distributions, thus supporting a large number of users for future IoT applications.

In Chap. 6, we have proposed a low-complexity DSL-based solution to reduce the complexity of the PSO-based solutions in Chap. 4. Similarly, we considered a UAV operating as a DF relay between BS and multiple obscured users, and HBF was designed for both UAV and the BS. The proposed DSL-based solution aimed to jointly optimize the UAV location and PA to users. Specifically, the DSL-based solution consisted of an offline SL phase and online UAV location and PA prediction. For this purpose, we train the SL model using PSO-based generated dataset, discussed in Chap. 4. The illustrative results depict that the proposed DSL-based solution closely approaches the PSO-based sum rate capacity with remarkably reduced runtime.

In Chap. 7, we have extended the work in Chap. 6, by introducing movement to IoT users, which is a realistic and close to real-world assumption. Considering HBF for both UAV and BS, a DRL algorithm, namely DDPG with continuous action space, is developed to train the UAV for its deployment. By using a customized reward function, the RL agent learns an optimal UAV deployment policy capable of adapting to both static and dynamic environments. The numerical results show that the proposed DDPG-based solution achieves the AR of its PSO-based counterpart, while greatly reducing the runtime in dynamic environments. Furthermore, the results reveal that the DSL-based solution fails in more complex, dynamic scenarios.

In Chap. 8, we have aimed to further reduce the complexity of the proposed DDPG-based UAV deployment in dynamic environments. For this purpose, we propose a novel DDPG codebook-based solution for the UAV deployment problem. The proposed codebook-based solution consists of an offline agent training phase and an online UAV deployment prediction to achieve the maximum capacity. The illustrative results show that the proposed DDPG codebook-based solution can closely approach the DDPG-based solution while with substantially reduced runtime. Additionally, the results show that deploying multiple UAVs can significantly increase the system's capacity to accommodate more users.

9.2 Future Works

This book has presented various AI-based solutions for UAV-assisted HBF mMIMO IoT systems. While concluding this thesis, we would like to point out possible avenues for the future research directions:

- **Integrated Wireless Information and Power Transfer (IWIPT):** It is expected that the number of smart devices will surpass 25.4 billion by 2030. With billions of low-power consumption devices, such as wireless sensors, connecting to the Internet through the IoT framework, there is a significant increase in both energy demand and data traffic. RF transmissions have long been utilized as a source of wireless information transfer (WIT). Recent advancements in energy harvesting (EH) circuit design have made RF signals a feasible source of power for wireless devices [1–3]. In contrast to traditional wireless networks, where nodes are powered by fixed energy sources, such as batteries, which must be replaced or replenished manually when they run out, RF-enabled wireless power transfer (WPT) eliminates the problem of battery replacement by ensuring a constant supply of energy. As a result, WPT is particularly appealing for applications in which replacing batteries is inconvenient (e.g., for IoT applications) or even impractical (e.g., for implanted devices in human body). In the context of this book, IWIPT can be explored for energizing UAVs, IoT nodes, or both. This includes designing adaptive EH mechanisms, managing interference effectively, and creating resource allocation algorithms that dynamically balance power and data needs. Additionally, conducting real-world experiments to validate these schemes would further solidify IWIPT's practical applications.

- **Joint Sensing and Communications (JSAC):** JSAC is a promising area that merges data transmission with environmental sensing, providing a unified platform for both communication and situational awareness. This convergence is crucial for applications such as autonomous vehicles, smart cities, and advanced surveillance systems, where real-time data and environmental context are essential [4, 5]. The HBF solutions presented in this thesis can be extended to simultaneously optimize communication and sensing tasks. This includes designing beamforming strategies that adapt based on real-time sensing data, integrating multiple sensing modalities, and exploring UAV-assisted sensing to enhance coverage and accuracy. Additionally, implementing these techniques in practical hardware and ensuring real-time adaptability in dynamic environments would significantly advance JSAC applications.
  Neat-Field Communications (NFC): Recently, NFC has gained significant attention as it possess the potential for high-capacity, low-latency communications over short distances. This is particularly important in densely populated environments, IoT networks, and UAV-assisted communications systems, where efficient and secure data transfer is crucial [6–8]. The research work of this book can be extended to focus on developing UAV-assisted HBF techniques tailored for NFC, optimizing signal focusing and spatial efficiency. This includes exploring UAV-assisted NFC applications in developing methods to mitigate interference, designing energy-efficient solutions, and integrating NFC in IoT devices for improved communication between UAVs and IoT nodes.
- **Multi-Agent Deep Reinforcement Learning (MA-DRL):** single-agent deep reinforcement learning (SA-DRL) methods have shown to be effective in solving problems addressed in this book. However, as the state or action space grows larger, SA-DRL struggles to maintain good performance. In such cases, MA-DRL can be a viable alternative, where by using training multiple agents, MA-DRL can tackle more complex challenges [9, 10]. Within this book, MA-DRL can be applied to solve the joint optimization problem of UAV location and PA discussed in Chap. 7. Specifically, as the number of users increases, SA-DRL may fail to find the best solution. Using MA-DRL, this joint optimization problem can be effectively addressed.
- **Semantic Communication and Generative AI:** Semantic communication is a new approach proposed for future communications systems [11, 12]. Instead of sending raw data, it focuses on transmitting the meaning of the message. This can help go beyond the traditional Shannon capacity by avoiding the need to send unnecessary information. Semantic communication is made possible through semantic encoding and decoding, which are supported by generative AI models. In the context of this book, semantic communication can be used to send information from the BS to the users or gateway more efficiently. While this approach can increase the system capacity, it may also introduce extra complexity due to the advanced processing it requires.

References

1. S. Özyurt, A. F. Coşkun, S. Büyükçorak, G. Karabulut Kurt, O. Kucur, A survey on multiuser SWIPT Communications for 5G+. IEEE Access **10**, 109814–109849 (2022)
2. A. Costanzo, D. Masotti, G. Paolini, D. Schreurs, Evolution of SWIPT for the IoT world: Near- and far-field solutions for simultaneous wireless information and power transfer. IEEE Microw. Mag. **22**(12), 48–59 (2021)
3. W. Feng et al., UAV-enabled SWIPT in IoT networks for emergency communications. IEEE Wirel. Commun. **27**(5), 140–147 (2020)
4. X. Fang, W. Feng, Y. Chen, N. Ge, Y. Zhang, Joint communication and sensing toward 6G: Models and potential of using MIMO. IEEE Internet Things J. **10**(5), 4093–4116 (2023)
5. A. Liu et al., A survey on fundamental limits of integrated sensing and communication. IEEE Commun. Surv. Tutor. **24**(2), 994–1034 (2022)
6. M. Cui, Z. Wu, Y. Lu, X. Wei, L. Dai, Near-field MIMO communications for 6G: fundamentals, challenges, potentials, and future directions. IEEE Commun. Mag. **61**(1), 40–46 (2023)
7. Y. Liu, Z. Wang, J. Xu, C. Ouyang, X. Mu, R. Schober, Near-field communications: a tutorial review. IEEE Open J. Commun. Soc. **4**, 1999–2049 (2023)
8. H. Zhang, N. Shlezinger, F. Guidi, D. Dardari, M.F. Imani, Y.C. Eldar, Beam focusing for near-field multiuser MIMO communications. IEEE Trans. Wirel. Commun. **21**(9), 7476–7490 (2022)
9. J. Cui, Y. Liu, A. Nallanathan, Multi-agent reinforcement learning-based resource allocation for UAV networks. IEEE Trans. Wirel. Commun. **19**(2), 729–743 (2020)
10. Z. Xia et al., Multi-agent reinforcement learning aided intelligent UAV swarm for target tracking. IEEE Trans. Veh. Technol. **71**(1), 931–945 (2022)
11. W. Yang et al., Semantic communications for future internet: fundamentals, applications, and challenges. IEEE Commun. Surv. Tuts. **25**(1), 213–250 (2023)
12. Q. Wu, H. Chen, Z. Yang, T.Q.S. Quek, H.V. Poor, Semantic communications: principles and challenges. IEEE Commun. Surv. Tuts. **23**(3), 1577–1604 (2021)

MIX
Papier aus verantwortungsvollen Quellen
Paper from responsible sources
FSC® C105338

If you have any concerns about our products,
you can contact us on
ProductSafety@springernature.com

In case Publisher is established outside the EU,
the EU authorized representative is:
Springer Nature Customer Service Center GmbH
Europaplatz 3, 69115 Heidelberg, Germany

Printed by Libri Plureos GmbH
in Hamburg, Germany